The Pocket Handbook of Image Processing Algorithms in C

AF684105

Harley R. Myler
Arthur R. Weeks

Department of Electrical & Computer Engineering
University of Central Florida
Orlando, Florida

Prentice Hall P T R
Englewood Cliffs, New Jersey 07632

Library of Congress Cataloging-in-Publication Data

Myler, Harley R.,
The pocket handbook of image processing algorithms in C/Harley R. Myler, Arthur R. Weeks.
p. cm.
Includes bibliographical references and index.
ISBN 0-13-642240-3
1. C (Computer program language) 2. Computer algorithms. 3. Image processing--Digital techniques.
I. Weeks, Arthur R. II. Title.
QA76.73.C15M95 1993
621.36'7'0285533-dc20 93-7791
CIP

Editorial production/supervision: *Harriet Tellem*
Cover design: *Karen Marsilio*
Buyer: *Mary E. McCartney*
Acquisitions editor: *Karen Gettman*

© 1993 by Prentice Hall P T R
Prentice-Hall, Inc.
A Simon & Schuster Company
Englewood Cliffs, New Jersey 07632

GIF (Graphic Interchange Format) is copyrighted/trademarked by Compuserve, Inc., TIFF (Tagged Interchange File Format) was developed by Microsoft Corporation and Aldus Corporation, PCX is a trademark of Z-Soft Corporation, Microsoft and MS-DOS are trademarks of Microsoft Corporation. MacPaint and Macintosh are trademarks of Apple Computer.

All rights reserved. No part of this book may be reproduced, in any form or by any means without permission in writing from the publisher.

Printed in the United States of America
10 9 8 7 6 5 4

ISBN 0-13-642240-3

90000

9 780136 422402

Prentice-Hall International (UK) Limited, *London*
Prentice-Hall of Australia Pty. Limited, *Sydney*
Prentice-Hall Canada Inc., *Toronto*
Prentice-Hall Hispanoamericana, S.A., *Mexico*
Prentice-Hall of India Private Limited, *New Delhi*
Prentice-Hall of Japan, Inc., *Tokyo*
Simon & Schuster Asia Pte. Ltd., *Singapore*
Editora Prentice-Hall do Brasil, Ltda., *Rio de Janeiro*

To Nancy, Krifka and Logan

To Dian, Anne, Frank, Anne and Richard

Table of Contents

This book is the culmination of an idea that was not ours uniquely, but one that has probably occurred to anyone who has spent long hours at the terminal writing Image Processing programs and routines. The programmer is often found surrounded by a library of texts, loaded with formulas and mathematically described algorithms. It is then necessary to engage in the arduous task of writing computer code from the mathematical description. In addition, if one does not have an adequate background in signal processing, the difficulty of this task is further compounded.

Programmers often spend time flipping from book to book, index to index, as they search for a useful mask or filter technique. In this book we have tried to bring within a single, desktop friendly cover--in a simple, easy to use format--the most popular of the vast ensemble of image processing algorithms. We chose the C programming language as our algorithm description format because of it's popularity, portability and widespread use in image processing systems. Also, the C language is compact and easy to learn, and a plethora of texts and tutorials exist in the literature for both the novice and seasoned programmer. Appendix B contains the complete code for a simple image processing program that may be used as a template for implementing any of the code in this handbook.

It is our hope that all programmers (in the end, are we not all programmers?) of Image Processing systems will find this book useful--both as a reference guide and for the implicit tutorial nature of our descriptions. Students should find our simple and direct discussion effective as a supplement to their classroom text or tutorial guides. Seasoned users will find the format convenient to keep close to the terminal for quick reference and easy access to definitions and algorithms.

Harley R. Myler
Arthur R. Weeks

How to Use This Book

TOPICAL LAYOUT

This book is intended to serve as a ready reference and imaging system user and developer companion. Algorithms and definitions covered in this book are arranged alphabetically and typically occupy one or two pages. Each algorithm follows a similar format that is illustrated by the miniaturized sample page graphic to the left, below.

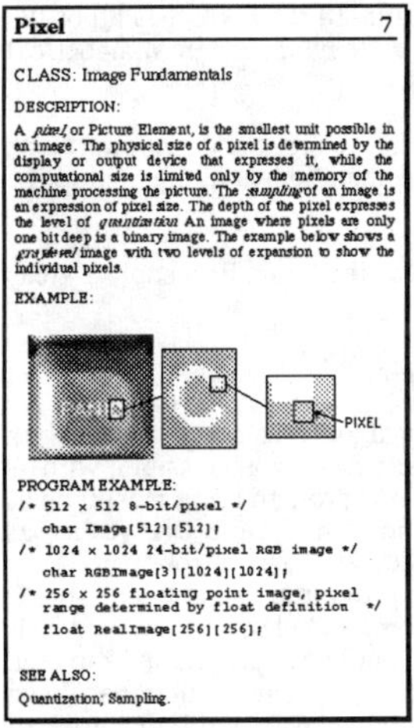

Pixel 7

CLASS: Image Fundamentals

DESCRIPTION:

A *pixel*, or Picture Element, is the smallest unit possible in an image. The physical size of a pixel is determined by the display or output device that expresses it, while the computational size is limited only by the memory of the machine processing the picture. The *sampling* of an image is an expression of pixel size. The depth of the pixel expresses the level of *quantization*. An image where pixels are only one bit deep is a binary image. The example below shows a *grayscale* image with two levels of expansion to show the individual pixels.

EXAMPLE:

PROGRAM EXAMPLE:

```
/* 512 x 512 8-bit/pixel */
   char Image[512][512];
/* 1024 x 1024 24-bit/pixel RGB image */
   char RGBImage[3][1024][1024];
/* 256 x 256 floating point image, pixel
   range determined by float definition  */
   float RealImage[256][256];
```

SEE ALSO:

Quantization; Sampling.

Each topic is identified by name in bold print in the header region of each page with the page number. The class of algorithms to which the topic belongs is identified next. There are fifteen classes, which are:

- Adaptive Filters
- Coding and Compression
- Color
- Graphics
- Histogram Operations
- Image Fundamentals
- Mensuration
- Morphological Filters
- Noise
- Nonlinear Filters
- Segmentation
- Spatial Filters
- Spatial Frequency Filters
- Storage Formats
- Transforms

Each class is described on a separate page and appears in alphabetic order with the topics. Classes are distinguished from topics by a bullet (•) in front of the class name in the page header.

A description of the algorithm or the definition of the topic follows the class designation. This is followed by an example of the algorithm or description discussed. Examples may be graphic illustrations or images. The algorithm in C is then given, if appropriate to the topic, followed by suggested follow-on topics under the SEE ALSO heading.

FINDING TOPICS

Well-known topics, such as *thresholding*, may be looked up rapidly by thumbing through the headers at the top of each page. If the class of an algorithm is known, you may go directly to the class description, which will list all the topics that fall under that class.

The Table of Contents in the front of the book lists all of the topics and class descriptions as they appear in alphabetical order with their page numbers.

The Class Groupings Index, following this discussion, lists the classes alphabetically and their topics along with page numbers.

The Subject Index, at the end of the book, lists subject areas and page numbers.

ALGORITHM PROGRAMMING

The C programming language algorithms given in this book have been written in as simple a fashion as possible, with as little dependence on sophisticated programming technique as possible. In the words of Einstein, "make everything as simple as possible, but not simpler." It is a relatively easy matter for the experienced programmer to make the C routines presented more efficient and elegant; however, it is not a simple matter for an inexperienced programmer to read and understand, i.e., simplify, a program that has been written in a complex and obtuse way with the goal of computational efficiency.

Throughout the book an image data structure has been used that allows flexibility while maintaining clarity. Image representation within a computer is best expressed as a rectangular array; however, when dealing with more than one pixel data type across a set of algorithms, a C data structure is the best solution for maintaining consistency throughout the algorithm ensemble. The data structure used here contains the rectangular size of the image, the type of data each pixel in the image represents, and a pointer to the image data itself.

Another consideration in the use of a data structure are the differences that exist between hardware platforms. Images, by nature, are large data objects. The structure that we use assumes that the user has allocated a data space large enough to accommodate the image being processed within any restrictive boundaries that may exist on the hardware being used.

The example graphic below shows our data structure and its relationship to an image.

```
Struct Image{
    int   Rows;
    int   Cols;
    unsigned char *Data;
    unsigned char  Type;
};
```

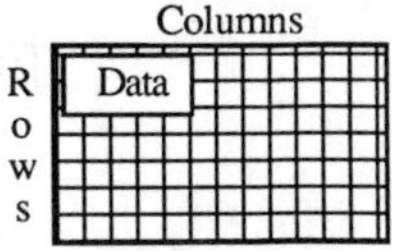

The **Data* pointer is a contiguous sequence of bytes whose type is described by the *Type* variable. The following four types of image data pixels are used:

Type = 0	Unsigned Character (BASIC)
Type = 2	Integer (UINT)
Type = 4	Float (REAL)
Type = 8	Complex (CMPLX)

For example, a rectangular image with Rows = 64 and Columns = 256 and a *Type* = 0 will require that 16,384 (256 × 64) bytes of storage be allocated for the *Data pointer. If *Type* = 2 and if an integer on the host machine requires 4 bytes, then four times the previous amount will have to be allocated.

One must also apply special data type modifiers when programming MS-DOS™ systems. The *far* and *huge* keywords must precede large data space pointers. Proper use of these modifiers is discussed further in Appendix B.

A complex image is a special case. Here, each datum represents a real and imaginary component using a float data type. If the host computer requires 4 bytes to represent a float value, then each datum in the *Data buffer will consist of 8 bytes: the first 4 bytes for the real part of the datum, and the second 4 bytes for the imaginary part.

Consider the diagram below as representing an image, where each pixel is represented by a square and the total number of pixels is unspecified. The pixels have been lettered to facilitate the discussion. The first row of the image consists of the set of pixels {A, B, ..., C}, the second row is {D, E, ..., F}, and the last row is {G, H, ..., I}. We can easily generalize this to sets of columns, as well.

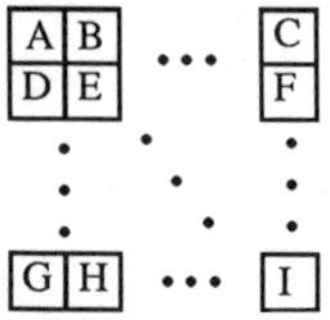

Using our image data structure and assuming pixels of Type 0, the arrangement of data for this image in memory would be as follows, where each pixel requires one byte of storage:

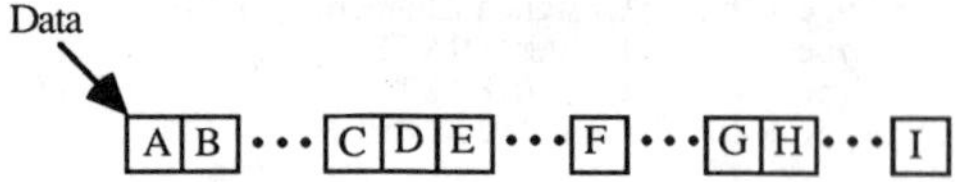

Data will point to a series of bytes that represent the pixel values in column-row order. If Type = 2 and an integer in our machine requires two bytes of storage (often this is the size of an *int*) then the internal representation will be:

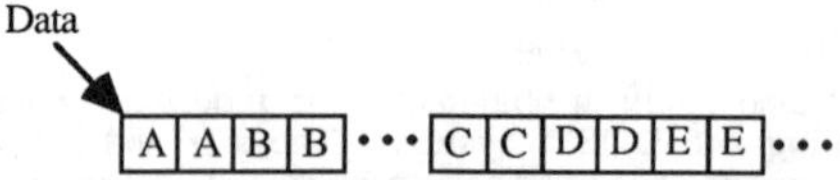

So pixel 'A' requires 2 bytes of storage, and so on. If the image is complex (Type = 8) and a float in the machine requires 4 bytes, the memory map will be:

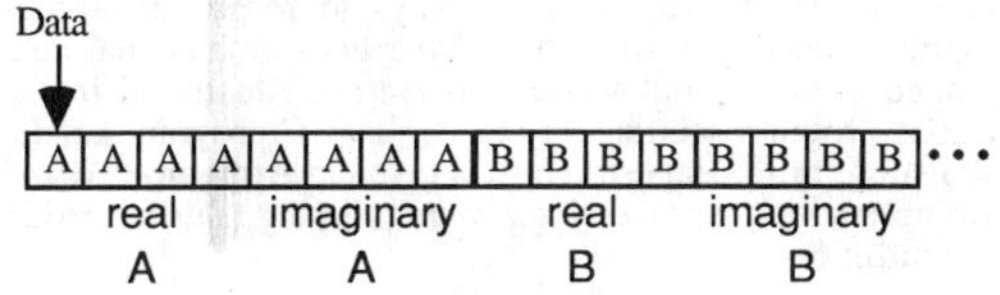

A complicity in programming arises in the accessing of the data using cartesian coordinates. The data structure allows for easy representation of a large number of pixel data formats and conforms readily to the idea of memory as a contiguous sequence of bytes; however, access of the data by coordinate is tricky. We must convert from the cartesian coordinate pair into a single value so that we may index into the linear image data array.

Consider the diagram below. We can visualize the pixels of the small 4 × 3 image to the left , with pixel values A through L, in terms of their row-column coordinates. The programming description could be specified as the array, **Img**. Pixel G would then be **Img[1][2]**. Alternatively, we can specify an image structure as described earlier, with the C statement **struct Image Img**. To access pixel G using our structure, we have to index into the data vector one row and three columns. The data in either case is stored as a sequence of bytes. The Data pointer of the image structure will be initialized to point to the start of this sequence, as shown in the graphic to the right, after the space has been allocated.

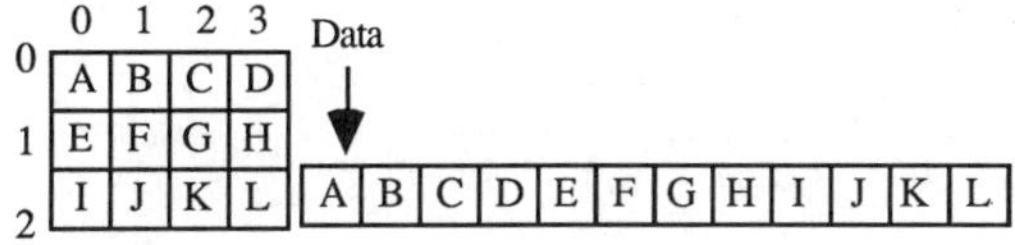

Using the structure format, the value of pixel G would be:

```
*(Img.Data +   (R * Img.Cols)   + C)
```

Where R is the row number (indexed from zero) and C is the column. For this example, *(Img.Data + 1*4 + 2) will yield the value, G. There are other ways to formulate access schemes into linear arrays that have two-dimensional data mapped to them, and some of these are illustrated in the various program sections of the book. A C program header file listing for the data structure is given in **Appendix A**, and a complete image processing program listing is described in **Appendix B**.

The routines presented in this book assume different image pixel data formats and the user must observe some caution when incorporating the algorithms in their own programs. The most popular pixel size is, of course, the unsigned character. For this reason, it was chosen as the base pointer data type in our image structure. Image transforms such as the Fourier and filters that operate in the spatial frequency domain use complex pixels. Other processes use float pixels. We have indicated the type of pixel data required by the algorithm in the description, but in the interest of compactness and simplicity have put no error detection mechanisms in the code. It is assumed that the user has allocated the correct data space for the algorithm called.

To simplify this concept of variable data widths for image pixels, we have provided an example of a multiple data format image I/O routine at the end of **Appendix B**.

Finally, a **Glossary** has been included with a large range of image processing terms defined as a convenience to the reader.

Class Groupings Index

This index is for reference by class membership. The classes and page numbers are:

• Color Image Processing Class

• Graphics Algorithms Class

• Histogram Techniques Class

Class Groupings Index

• Image Fundamentals Class

• Mensuration Class

• Morphological Filters Class

• Morphological Filters Class (continued)

• Noise Class

• Nonlinear Filters Class

• Segmentation Class

• Spatial Filters Class

• Spatial Frequency Filters Class

• Storage Formats Class

• Transforms Class

CLASS: Adaptive Filters

DESCRIPTION:

The adaptive double window modified trimmed mean (DW-MTM) filter overcomes the difficulties of using the MMSE filter in the presence of impulsive noise by using the median estimator to estimate the local mean. A new local mean is then computed using only pixels within a small graylevel range about the median. This effectively removes outliers in the calculation of the mean estimate, hence improving the overall performance of the mean filter in the presence of outliers such as salt and pepper noise.

The adaptive DW-MTM filter algorithm is described as follows. Given a pixel located at x, y within the image, a median filter (MED[g(x, y)]) is computed within an $n \times n$ local region surrounding the location x, y. The median value computed from this filter is used to estimate the mean value of the $n \times n$ local area. Next, a larger-sized window surrounding the pixel at location x, y of size $q \times q$ is used to calculate the mean value. In computing the mean value in the $q \times q$ window, only pixels within the graylevel range of

$$\mathrm{MED}[g(x, y)] - c \text{ to } \mathrm{MED}[g(x, y)] + c$$

are used, eliminating any outliers from the mean calculation. The output of the DW-MTM filter is the $q \times q$ mean filter. The value of c is chosen as a function of the noise standard deviation as

$$c = K \cdot \sigma_n$$

Typical values of K range from 1.5 to 2.5. This range for K is based on the assumption that for Gaussian noise statistics the peak-to-peak graylevel variations will be in the range of $\pm 2\sigma_n$ 95% of the time and any values outside this range are more than likely outliers. For K = 0, the DW-MTM filter reduces to a $n \times n$ median filter, and for K very large, the DW-MTM reduces to a $q \times q$ mean filter. Hence, as K decreases, the filter does a better job of filtering impulsive noise, but does a poor job of filtering uniform and Gaussian-type noise.

EXAMPLE:

(a) *(b)*

(a) The original Gaussian (variance = 200) and salt and pepper noise (probability = 10%) corrupted image and *(b)* the DW-MTM filtered image with n = 3, q = 5, K=1.5, and σ_n = 40.963.

ALGORITHM:

The program assumes that the original image is a 256 graylevel × IMAGE->Rows × IMAGE->Cols pixel image stored in the structure IMAGE. The program also assumes that the standard deviation of the noise and the threshold parameter K is passed to the program upon execution. The program first computes the median in a 3 × 3 local window. This median value is then used to determine which points are excluded in the 5 × 5 mean filter calculation. Upon completion of the program, the filtered image is stored in the structure IMAGE1.

```
DWMTM_filter(struct Image *IMAGE, struct
Image *IMAGE1,  float NSTD, float K)
      {
      int X, Y, x1, y1, med[9];
      int median;
      int gray, i, j, temp;
      long int total, sum, R;
      R=IMAGE->Cols;
      for(Y= 2; Y<IMAGE->Rows-2; Y++)
         for(X=2; X<IMAGE->Cols-2; X++)
            {
```

```
        total=0;
        for(y1=-1; y1<=1; y1++)
           for(x1=-1; x1<=1; x1++)
              {
              med[total]=
              *(IMAGE->Data+
              X+x1+(long)(Y+y1)*R);
              total=total+1;
              }
        for(j=1; j<=8;j++)
           {
           temp = med[j];
           i=j-1;
           while(i>=0 && med[i] >temp)
              {
              med[i+1]= med[i];
              i=i-1;
              }
              med[i+1]=temp;
           }
        median=med[4];
        sum=0; total=0;
        for(y1=-2; y1<=2; y1++)
           for(x1=-2; x1<=2; x1++)
              {
              gray= *(IMAGE->Data
              +X+x1+(long)(Y+y1)*R);
              if(gray>=(median-K*NSTD))
                if(gray<=(median+K*NSTD))
                  {
                  sum=sum+gray;
                  total = total +1;
                  }
              }
        *(IMAGE1->Data+X+(long)Y*R)
        =(unsigned char)
        ((float)sum/(float)total);
        }
  }
```

SEE ALSO: Adaptive MMSE Filter

Adaptive Filters

DESCRIPTION:

Depending on the type of noise that is present within an image, the type of filter that is chosen to remove the noise can radically affect the important details that must remain unaffected by the filtering operation. For example, *Gaussian* type noise is better filtered using a mean filter while *Salt and Pepper* type noise is better filtered using a median filter. The disadvantage of using a mean filter over a median filter is that the mean filter removes high spatial frequencies which blurs the edges within an image.

An ideal filter to use on an image is a filter that adaptively changes its filtering characteristics depending on the image content within a local window. For example, if the image content within a local window contains only edges, then a median filter would be used to preserve the edge details. If, within a local window, a uniform background is detected, then the filter would change its characteristics to perform a mean filter within this local window. Filters that dynamically change their filtering characteristics as they are scanned through the image are known as adaptive filters.

CLASS MEMBERSHIP:

DW-MTM Filter

MMSE Filter

SEE ALSO: Noise, Nonlinear and Spatial Filters

Adaptive MMSE Filter

CLASS: Adaptive Filters

DESCRIPTION:

The Minimum Mean-Square Error filter (MMSE) makes use of the knowledge of the local variance to determine if a mean filter is to be applied to the local region of an image. This filter works best for additive type short tail noise. The output of the MMSE filter for a pixel located at the coordinates x, y in an image g(x, y) is

$$r(x, y) = \left(1 - \frac{\sigma_n^2}{\sigma_l^2}\right) \cdot g(x, y) + \frac{\sigma_n^2}{\sigma_l^2} \cdot K ,$$

where σ_n^2 is the variance of the noise, and σ_l^2 is the local variance about the pixel located at x, y.

The parameter K is the output from a local mean filter which is usually a 3 × 3 or 5 × 5 mean filter. In background regions the local variance is approximately equal to the noise variance reducing the adaptive MMSE to approximately a local mean filter. Near edges, the local variance will be much larger than the noise variance and the adaptive MMSE filter reduces to the original unfiltered image.

EXAMPLE:

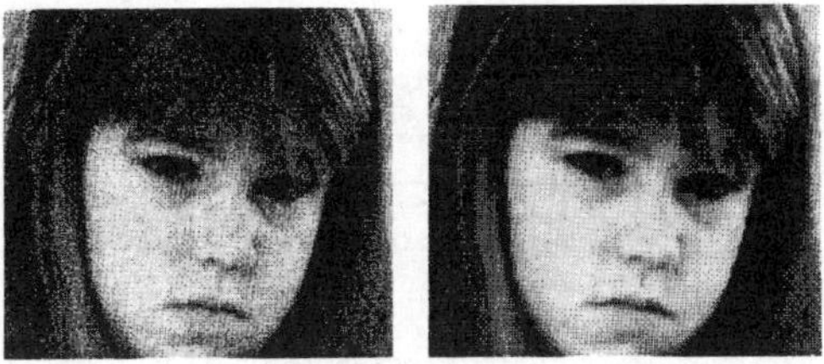

(a) *(b)*

(a) The original additive Gaussian noise corrupted image (variance = 200) and ***(b)*** the 5 × 5 MMSE filtered image.

ALGORITHM:

The program assumes that the original image is a 256 graylevel × IMAGE->Rows × IMAGE->Cols pixel image stored in the structure IMAGE. The program also assumes that the variance of the noise is known and is passed to the program upon execution. The program first computes the local variance over a N × N window from the local first and second moments. Next, the program computes the MMSE filter output using the N × N mean computed in the first step. Upon completion of the program, the filtered image is stored in the structure IMAGE1.

```
MMSE_filter(struct Image *IMAGE, struct
Image *IMAGE1, float NVAR)
      {
      int X, Y, x1, y1, N, g;
      long int total, sum, sum1, R;
      float FSECOND, FVAR, FMEAN;
      R=IMAGE->Cols;
      N=5;
      for(Y= N/2; Y<IMAGE->Rows-N/2; Y++)
         for(X=N/2;X<IMAGE->Cols-N/2; X++)
            {
            sum=0;
            sum1=0;
            total =0;
            for(y1=-N/2; y1<=N/2; y1++)
               {
               for(x1=-N/2; x1<=N/2; x1++)
                 {
                 sum=sum + *(IMAGE->Data +
                 X+x1+(long)(Y+y1)*R);
                 sum1=sum1 +
                 (long)*(IMAGE->Data +
                 X+x1+(long)(Y+y1)*R)*
                 (long)*(IMAGE->Data +
                 X+x1+(long)(Y+y1)*R);
                 total = total +1;
                 }
               }
              FSECOND=(float)sum1/
              (float) total;
              FMEAN = (float)sum/
              (float)total;
              FVAR=FSECOND - FMEAN*FMEAN;
            if(FVAR ==0.0)
```

```
          g=(int)(FMEAN+.5);
        else
          g=(int)((1-NVAR/FVAR)* *(
          IMAGE->Data+X+(long)Y*R) +
          NVAR/FVAR*FMEAN+.5);
        if(g>255)
          g=255;
        if(g<0)
          g=0;
        *(IMAGE1->Data+X+(long)Y*R)=g;
        }
  }
```

SEE ALSO: Adaptive DW-MTM Filter

Alpha-Trimmed Mean Filter

CLASS: Nonlinear Filters

DESCRIPTION:

The alpha-trimmed mean filter is based on order statistics and varies between a median and a mean filter. It is used when an image contains both short and long tailed types of noise. For example, many images contain both Gaussian and salt and pepper type noise. To define the alpha-trimmed mean filter, all pixels surrounding the pixel at the coordinate x, y in the image *A* which are specified by an input mask A(i) are order from minimum to maximum graylevel value

$$A_1 \leq A_2 \leq A_3 \leq A_4 \leq \cdots \leq A_{N-1} \leq A_N \,.$$

The alpha-trimmed mean filter is then given as

$$\text{AlphaMean}\,(A) = \frac{1}{N-2P} \sum_{i=P}^{N-P} A_i \,,$$

where N − 2P is the total number of pixels included in the average calculation.

EXAMPLE:

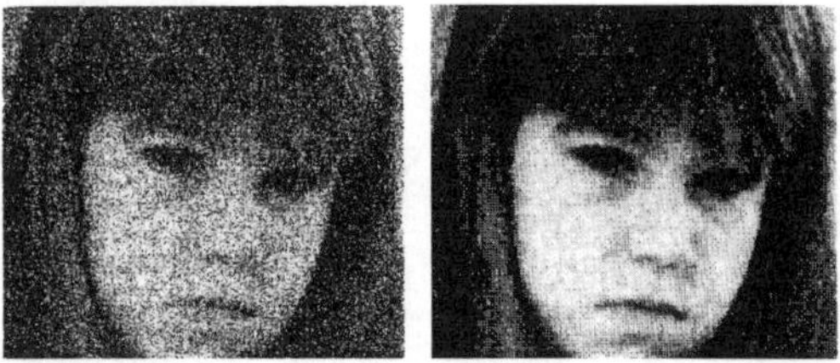

(a) *(b)*

(a) A Gaussian and salt and pepper noise corrupted image and ***(b)*** the alpha-trimmed mean filtered image using a 5 × 5 square mask and P = 3.

ALGORITHM:

The program assumes that the original image is a 256 graylevel × IMAGE->Rows × IMAGE->Cols pixel image stored in the structure IMAGE. The program computes the alpha-trimmed mean filter over a set of pixels contained within a square N × N region of the image centered at the pixel X, Y. The size of the filtering operation is determined by the variable N and should be set to an odd number and be less than 12. The parameter P passed to the program determines how many of the endpoint pixels are eliminated from the ordered data. For example, if P = 1, the pixels with the minimum and maximum graylevel values are not included in the average calculation. For the special cases of P = 0 and P= (N * N) / 2 - 1/2 the alpha-trimmed mean filter becomes the mean and median filters respectively. Upon completion of the program, the alpha-trimmed mean filtered image is stored in the structure IMAGE1.

```
AlphaMean(struct Image *IMAGE, int P,
struct Image *IMAGE1)
      {
      int X, Y, I, J, SUM, Z;
      int N, AR[121], A;
      N=7;
      for(Y=N/2; Y<IMAGE->Rows-N/2; Y++)
         for(X=N/2; X<IMAGE->Cols-N/2; X++)
            {
            Z=0;
            for(J=-N/2; J<=N/2; J++)
               for(I=-N/2; I<=N/2; I++)
                  {
                  AR[Z]=*(IMAGE->Data+X
                  +I+(long)(Y+J)
                  *IMAGE->Cols);
                  Z++;
                  }
            for(J=1; J<=N*N-1;J++)
               {
               A = AR[J];
               I=J-1;
               while(I>=0 && AR[I] >A)
                  {
                  AR[I+1]=AR[I];
                  I=I-1;
                  }
```

```
          AR[I+1]=A;
        }
      SUM=0;Z=0;
      for(J=P; J<=N*N-1-P;J++)
        {
        SUM = SUM + AR[J];
        Z++;
        }
        *(IMAGE1->Data+X+(long)Y
        *IMAGE->Cols) = (unsigned
        char)((float)SUM/(float)Z
        +.5);
      }
  }
```

SEE ALSO: Geometric, Y_p, Harmonic and Arithmetic Mean Filters, Median, and other Nonlinear Filters.

CLASS: Mensuration

DESCRIPTION:

Area algorithms measure the number of pixels contained within an object in an image. The boundary of the object must be defined, or the region wherein the object lies as well as the pixel value to be counted as part of the object must be known. In the image below, a region is delineated and a pair of objects lie within the region. The area algorithm given returns an area calculation for an object given a specific graylevel. When the physical size of a pixel is known, the actual area in size units can be calculated.

The example below shows the original image on the left. To the right is a thresholded, binarized, and eroded image that has a region drawn around a white object that represents the "end" key. The white pixels in this region, when counted, represent the area of the object.

EXAMPLE:

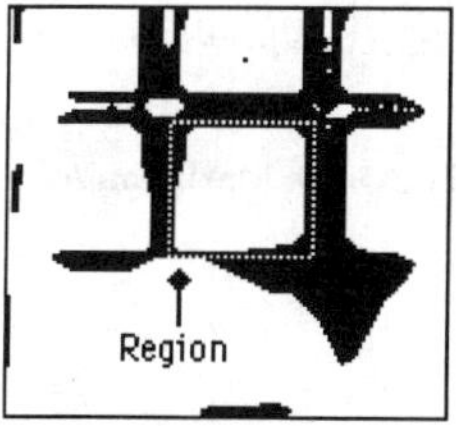

ALGORITHM:

The algorithm accepts an image structure pointer, **In**, that is assumed to be an 8-bit unsigned character picture. The coordinates of the region containing the object, **(x1,y1,x2,y2)**, and the object graylevel, **ObjVal**, are also passed. The area is computed by counting the number of pixels in the area at the graylevel value. This value is returned.

Objects that contain spaces or multiple objects within the region containing pixels at the object graylevel value are also counted. Other methods of area calculation can avoid these problems, but require the contour pixels of the object. If the contour is known, the area can be calculated line by line by counting candidate pixels between contour coordinates. The macro **pix** is used for simplified access to the pixels of the image by coordinates.

```
#define pix(Im,x,y) \
        *(Im->Data + (x)*Im->Cols + (y))

/* Compute and return area for objects */

int area(struct Image *In,int x1,int y1,
         int x2,int y2,unsigned char ObjVal)
{
   long i,j,rows;
   int area_value = 0;

   for(i=x1; i<=x2; ++i)
      for(j=y1; j<=y2; ++j){
         if(pix(In,i,j)==ObjVal)++area_value;
      }
   return(area_value);
}
```

SEE ALSO: Maximum Axis, Minimum Axis

CLASS: Spatial Filters

DESCRIPTION:

An arithmetic mean filter operation on an image removes short tailed noise such as uniform and Gaussian type noise from the image at the cost of blurring the image. The arithmetic mean filter is defined as the average of all pixels within a local region of an image. Pixels that are included in the averaging operation are specified by a mask. The larger the filtering mask becomes the more predominant the blurring becomes and less high spatial frequency detail that remains in the filtered image. The definition of an arithmetic mean filter in terms of an image A is

$$\text{Mean}\,(A) = \frac{1}{N^2} \sum_{(i,\,j)\,\in\,M} A(x+i,\; y+j)\,,$$

where the coordinate $x + i$, $y + j$ is defined over the image A and the coordinate i, j is defined over the mask M. The total number of pixels specified by the mask and included in the average operation is given by N^2.

EXAMPLE:

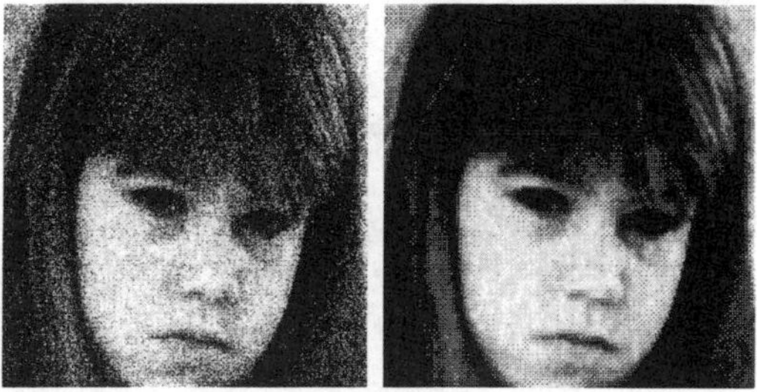

(a) *(b)*

(a) The original Gaussian noise corrupted image and ***(b)*** the mean filtered image using a 7×7 mask.

ALGORITHM:

The program assumes that the original image is a 256 graylevel × IMAGE->Rows × IMAGE->Cols pixel image stored in the structure IMAGE. The program then performs a N × N arithmetic mean filter on the image. The size of the filtering operation is determined by the variable N and should be set to an odd number. Upon completion of the program the filtered image is stored in the structure IMAGE1.

```
Mean(struct Image *IMAGE, struct Image
*IMAGE1)
      {
      int X, Y;
      int I, J;
                int N, SUM;
      N=7;
      for(Y=N/2; Y<IMAGE->Rows-N/2; Y++)
         {
         for(X=N/2; X<IMAGE->Cols-N/2; X++)
            {
            SUM=0;
            for(J=-N/2; J<=N/2; J++)
               {
               for(I=-N/2; I<=N/2; I++)
                  {
                  SUM=SUM + *(IMAGE->Data+X
                  +I+(long)(Y+J)
                  *IMAGE->Cols);
                  }
               }
            *(IMAGE1->Data+X+(long)Y
            *IMAGE->Cols) = SUM/N/N;
            }
         }
      }
```

SEE ALSO: Median, Minimum, Maximum, and other Nonlinear Filters

CLASS: Histogram Operation

DESCRIPTION:

Brightness correction is used to enhance the visual appearance of an image. Brightness modification of an image is defined as

$$s_i = g_i + \text{brightness} \ ,$$

where s_i is the ith graylevel value of the brightness enhanced image, and g_i is the ith graylevel value of the original image.

EXAMPLE:

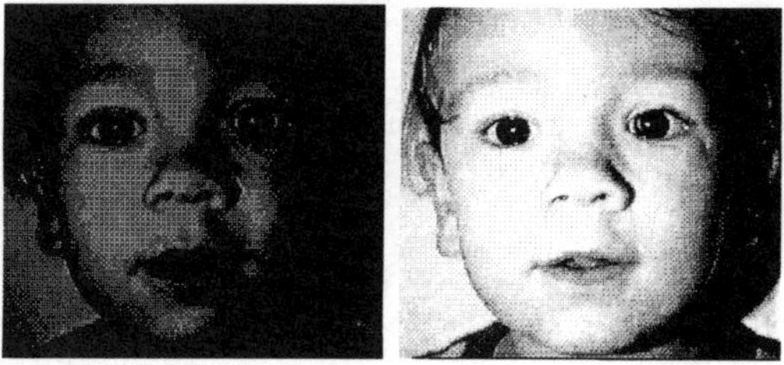

(a) ***(b)***

(a) The original image and ***(b)*** the brightness corrected image.

ALGORITHM:

The program assumes that the original image is a 256 graylevel × IMAGE->Rows × IMAGE->Cols pixel image stored in the structure IMAGE. The program adds the value passed to the program in the brightness parameter to each pixel in the image. Upon completion of the program, the brightness corrected image is stored in the structure IMAGE1.

```
Brightness(struct Image *IMAGE, struct
Image *IMAGE1,  int brightness)
     {
```

```
int X, Y, I;
for(Y=0; Y<IMAGE->Rows; Y++)
   {
   for(X=0; X<IMAGE->Cols; X++)
      {
      I= *(IMAGE->Data+X+(long)Y*
      IMAGE->Cols) + brightness;
      if(I> 255)
        I=255;
      if(I<0)
        I=0;
      *(IMAGE1->Data+X+(long)Y*
      IMAGE->Cols) = I;
      }
   }
}
```

SEE ALSO: Graylevel Histogram, Histogram Specification, Contrast Correction, and Nonlinear Graylevel Transformations

CLASS: Color Image Processing

DESCRIPTION:

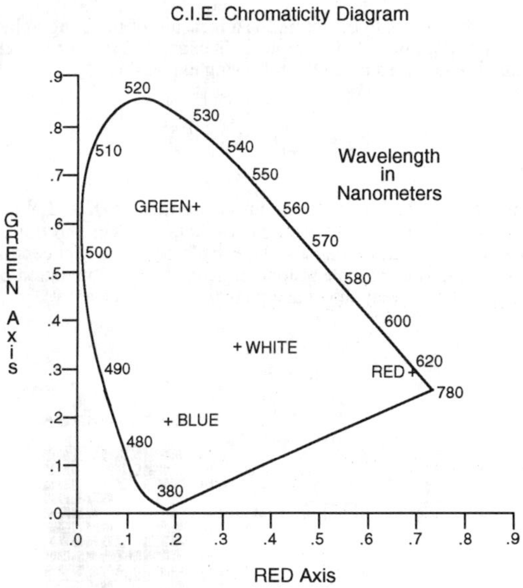

1 = Red + Green + Blue

or

Blue = 1- Red - Green

SEE ALSO: RGB and YIQ Color Models, Hue and Saturation Correction, and Pseudocolor

CLASS: Mensuration

DESCRIPTION:

Centroid refers to center and is a measure of the geographic center of an object. The centroid is expressed in coordinates and is calculated using the following expression:

$$X_c = \frac{1}{A}\sum_{i=1}^{N} X \quad Y_c = \frac{1}{A}\sum_{i=1}^{N} Y$$

where X_c and Y_c are the centroid coordinates, X and Y are the coordinates of the ith object pixel, and A is the object area. The centroid may also be calculated as the first central moment. The example graphic illustrates the computation of the centroid of an object using the formula given above.

EXAMPLE:

$$Xc = 39/13 = 3$$

$$Yc = 52/13 = 4$$

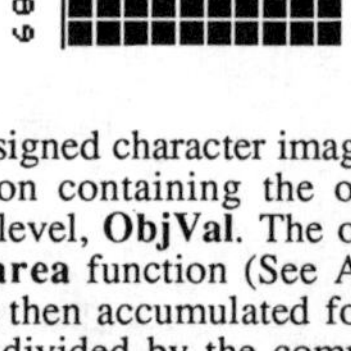

ALGORITHM:

The algorithm accepts an 8-bit unsigned character image, **In** and the coordinates of the region containing the object (**x1,y1,x2,y2**) and the object graylevel, **ObjVal**. The object area is computed by calling the **area** function (See Area). Row and column coordinates are then accumulated for the object. These accumulations are divided by the computed area for the object and the resulting centroid coordinates are returned as the **x** and **y** members of the **coord** data structure. Integer arithmetic assures that a coordinate value will be returned. If **area_value** computes to zero, negative coordinates are returned.

```
/* coordinates structure */
struct coord {
   float x,y;
};

/* Compute and return centroid */

void *centroid(struct Image *In,int x1,
               int y1,int x2,int y2,
               unsigned char ObjVal,
               struct coord *coords)
{
   long i,j;
   int area_value, Xcent=0, Ycent=0;

   area_value =
      area(In, x1, y1, x2, y2,ObjVal);

   if(area_value == 0){
      coords->x = -1; coords->y = -1;
      return;
   }

   for(i=x1; i<=x2; ++i)
      for(j=y1; j<=y2; ++j){
         if(pix(In,i,j)==ObjVal){
            Xcent += j;
            Ycent += i;
         }
      }

   coords->x = Xcent/area_value + 1;
   coords->y = Ycent/area_value + 1;

   return;
}
```

SEE ALSO: Area, Maximum Axis, Minimum Axis, Moments

CLASS: Coding and Compression

DESCRIPTION:

Chain Coding is used to define object boundaries for the purpose of measurement and identification. A chain code follows the boundary of an object and indicates the change in direction according to a 4- or 8-direction pattern as shown here:

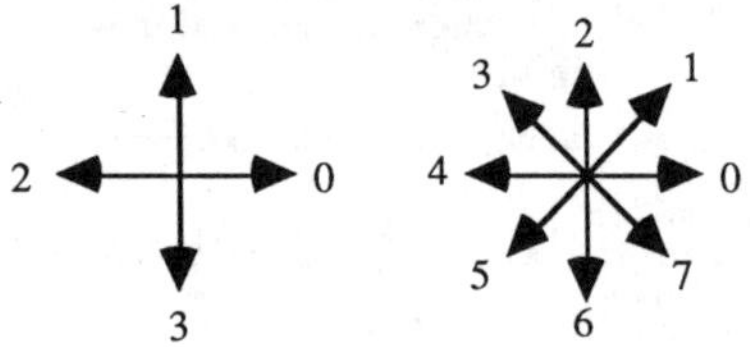

The code is dependent on where on the boundary the code sampling starts, but may be normalized using a number of techniques.

EXAMPLE:

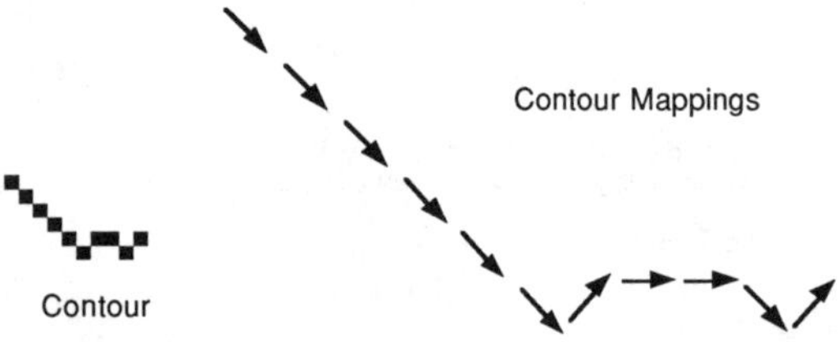

A contour is shown with the directional mappings given to the right. The 8-direction chain code for this contour would be:

```
77777710071.
```

ALGORITHM:

The routine is passed an image structure pointer of unsigned character data containing the open ended contour, **In**, the **x** and **y** coordinates of the start of the contour to be coded, and a threshold value, **thresh**, used to determine membership in the contour from a pixel neighborhood. The routine is also supplied a character string pointer, **code**, which must be initialized to the maximum expected length of the code. The routine returns the length of the code in the int pointer **length**. Caution must be exercised in that the code string will contain zeroes that indicate the right horizontal direction and not string termination.

The routine uses a macro, **idx**, to simplify coordinate access into the image array.

```
/* compute 8-direction chain code given
start point of contour */

#define idx(m,n)/
   *(In->Data+(m)+((n)*In->Cols))

chain_8(struct Image *In, int x, int y,
        char *code,int *length,
        unsigned char thresh)
{
   *length = 1;

   for(;;){

      idx(x,y) = 0;

      if(idx(x+1,y)>thresh){           /* -> */
         *(code++) = 0;
         ++x;
         ++(*length);
         continue;
      }

      if(idx(x+1,y-1)>thresh){         /* / */
         *(code++) = 1;
         --y; ++x;
         ++(*length);
         continue;
      }
```

```
      if(idx(x,y-1)>thresh){                    /* | */
         *(code++) = 2;
         --y;
         ++(*length);
         continue;
      }

      if(idx(x-1,y-1)>thresh){                  /* \ */
         *(code++) = 3;
         --y;--x;
         ++(*length);
         continue;
      }

      if(idx(x-1,y)>thresh){                    /* <- */
         *(code++) = 4;
         --x;
         ++(*length);
         continue;
      }

      if(idx(x-1,y+1)>thresh){                  /* / */
         *(code++) = 5;
         ++y;--x;
         ++(*length);
         continue;
      }

      if(idx(x,y+1)>thresh){                    /* | */
         *(code++) = 6;
         ++y;
         ++(*length);
         continue;
      }

      if(idx(x+1,y+1)>thresh){                  /* \ */
         *(code++) = 7;
         ++y;++x;
         ++(*length);
         continue;
      }
      return;
   }
}
```

SEE ALSO: Mensuration

CLASS: Mensuration

DESCRIPTION:

Circularity is the measure of an object's closeness to circular shape. It can be expressed as the variance of the boundary pixel's distance to the centroid. If the object is perfectly circular, this variance will be zero. The limit on circularity is one-half the major axis.

ALGORITHM:

The circularity algorithm requires both the boundary pixels of the object in question as well as the centroid. The boundary may be determined by subtracting the single pixel erosion (See Erosion) of the object from the object. The algorithm accepts an 8-bit unsigned character image, **In** and the coordinates of the region containing the object (**x1**, **y1**, **x2**, **y2**) boundary and the object graylevel, **ObjVal**. The object centroid is computed by calling the **centroid** function (See Centroid); then a euclidian distance measure is applied to each boundary pixel with the centroid. These values are accumulated and the variance calculated.

```
/* coordinates structure */
struct coord {
   float x,y;
};

/* Compute and return circularity */

double circularity(struct Image *In,long x1,
                   long y1,long x2,long y2,
                   unsigned char ObjVal)
{
   struct coord Ocenter;
   long i,j,rows,points=0;
   double mean=0.0,temp,stdev=0.0,variance;

   centroid(In,x1,y1,x2,y2,ObjVal,&Ocenter);

/* compute mean */
   for(i=x1; i<=x2; ++i)
      for(j=y1; j<=y2; ++j){
         if(pix(In,i,j)==ObjVal){
```

```
            mean += sqrt(
            (i-Ocenter.x)*(i-Ocenter.x)+
            (j-Ocenter.y)*(j-Ocenter.y));
            ++points;
         }
      }

      if(points==0)return(-1);
      mean /= (double)points;

/* compute variance */
   for(i=x1; i<=x2; ++i)
      for(j=y1; j<=y2; ++j){
         if(pix(In,i,j)==ObjVal){
            temp = sqrt((i-Ocenter.x) *
                        (i-Ocenter.x) +
                        (j-Ocenter.y) *
                        (j-Ocenter.y))- mean;
            stdev += temp*temp;
         }
      }
   stdev /= (double)points;
   variance = sqrt(stdev);
   return(mean/variance);
}
```

SEE ALSO: Centroid, Maximum Axis, Minimum Axis, Moments

Circularly Symmetric Filter

CLASS: Spatial Frequency Filters

DESCRIPTION:

The *Circularly Symmetric Filter* allows for powerful frequency domain filtering using the FFT or a sequency-based transform such as the Walsh or Slant. The filter begins with a single dimension function that is rotated about it's vertical axis to create a filter volume that is applied to the frequency spectra of an image. Consider the following function, H(u) and the rotation H(u,v):

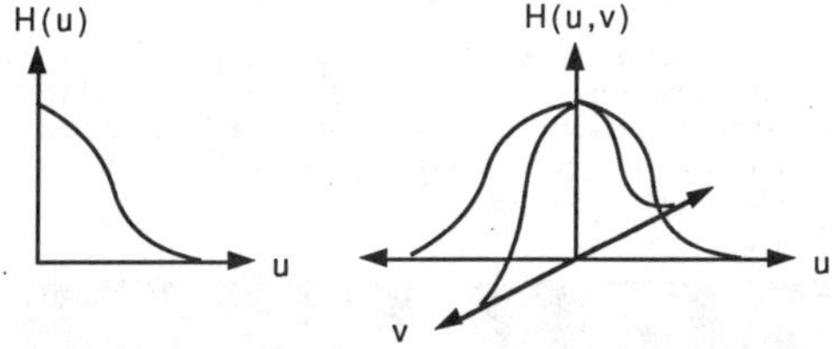

If the function H(u,v) as shown above is normalized to a maximum value of 1.0 and then multiplied with the frequency spectra of an image, F(u,v), the inverse FFT of the result will be a filtered picure. This assumes that the spectra has been centered in the spatial frequency domain. To center the spectra, each pixel in the original image can be multipled by $(-1)^{x+y}$ prior to taking the FFT. This takes advantage of the translation property of the Fourier. Otherwise, the filter origin must be shifted, as diagrammed below:

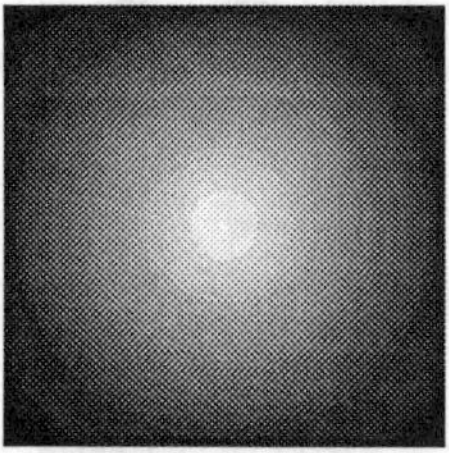

Image of Filter

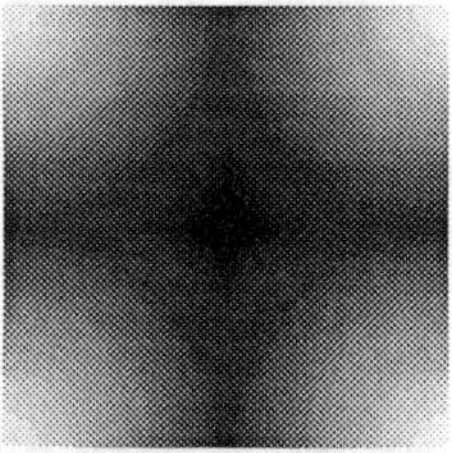

Filter after origin shifting

Filters can be generated by a number of techniques, the simplest of which is to assume an ideal filter and use a binary image as the filter function. This will most often yield undesirable results as the filtered image will display ringing, the subsampling artifacts that occur due to the convolution taking place. A convenient function for producing smooth transition filters is a two dimensional lowpass Butterworth, given by the following expression:

$$H(u,v) = \frac{1}{1 + \left[\frac{D_0}{\sqrt{u^2 + v^2}}\right]^{2n}}$$

The filter function shown on the previous page is derived from this equation. Other shapes are possible, including filters that selectively attentuate bands of frequencies or regions of the image. Inversion of the fraction in brackets will yield a high pass filter.

EXAMPLE:

Original Image

Fourier Magnitude (uncentered)

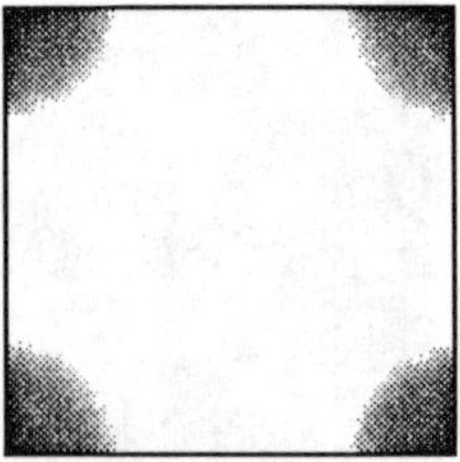

Magnitude after filter

Filtered (blurred) result

ALGORITHM:

The routine described, **circ_fil**, accepts a N × N Image structure pointer, **filt**, that is of type complex (sequential real and imaginary pixels of type float) and returns a *shifted* Butterworth lowpass filter function in the structure. N should be a power of 2 so that the filter may be multiplied with an image spectra. Filter parameters such as the spatial cutoff frequency, **Do**, and the filter order, **n**, are also passed. The filter shown in the discussion was generated by this function on a 128 × 128 image structure with n = 1 and Do = 64.

The filter is applied by multiplying each pixel of the complex spectra of an image by the filter image returned in **filt**. This may be accomplished using a simple pixel by pixel multiply routine. Note the multiply is the dot product of each image pixel real and imaginary part with each filter image real and imaginary part.

```
circ_filt(struct Image *filt,float Do,int n)
{
   int i,j,l,m,N;
   float *f,Huv;

   f = (float *)filt->Data;
   N = filt->Rows;

   for(i=0,l=N-1; i<N/2;++i,--l)
      for(j=0,m=N-1; j<N/2;++j,--m){
         /* low pass butterworth at (i,j) */
         Huv = 1.0/(1.0 +
               pow(sqrt(i*i+j*j)/Do,2*n));
         /*  <REAL>        <IMAGINARY>   */
         /* upper left quadrant */
         *(f+(i*N+j)) = *(f+(i*N+j)+1) = Huv;
         /* lower right quadrant */
         *(f+(l*N+m)) = *(f+(l*N+m)+1) = Huv;
         /* upper right quadrant */
         *(f+(i*N+m)) = *(f+(i*N+m)+1) = Huv;
         /* lower left quadrant */
         *(f+(l*N+j)) = *(f+(l*N+j)+1) = Huv;
      }
}
```

SEE ALSO: Fourier Transform

CLASS: Morphological Filters

DESCRIPTION:

Morphological closing of a binary object is defined as the dilation of that object followed by the erosion of the dilated object. The closing filter operation will reduce small inward bumps and small holes within the object..

$$\text{Close}(A, B) = (A \oplus B) \ominus B$$

EXAMPLES:

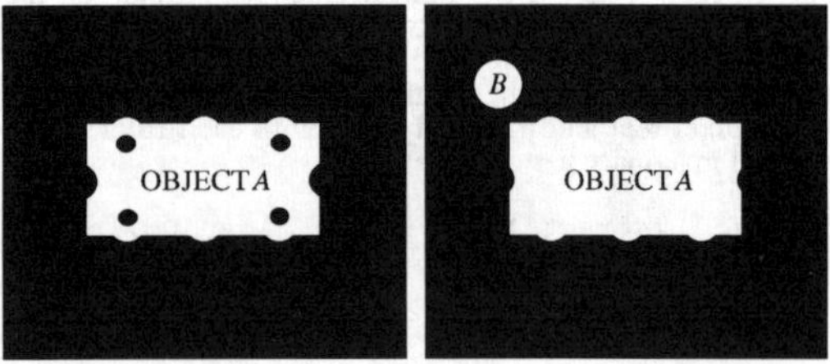

(a) *(b)*

(a) The original binary image of object A and ***(b)*** the closed image of object A with a circular structuring function B.

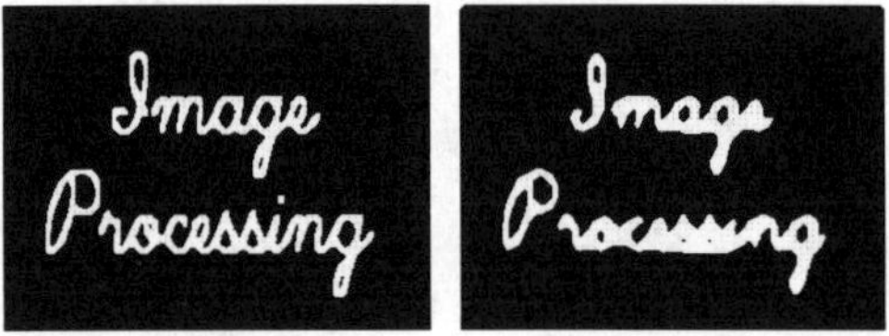

(a) *(b)*

(a) The original image and ***(b)*** the closed image.

ALGORITHM:

The program assumes that the original binary image is an IMAGE->Rows × IMAGE->Cols pixel image with a background graylevel value of 0 and an object graylevel value of 255 (object) and is stored in the structure IMAGE. The N × N structuring function is stored in array MASK[][]. Upon completion of the program, the closed image is stored in the structure FILTER. The binary erosion and dilation functions used by the algorithm can be found under binary erosion and dilation respectively.

```
#define N 5

Close(struct Image *IMAGE, int
   MASK[][N], struct Image *FILTER)
      {
      int X, Y;
      Dilation(IMAGE, MASK, FILTER);
      for(Y=0; Y<IMAGE->Rows; Y++)
         {
         for(X=0; X<IMAGE->Cols; X++)
            {
            *(IMAGE->Data+ X +
            (long)Y *IMAGE-> Cols) =
            *(FILTER->Data+ X + (long)Y
            * IMAGE->Cols);
            }
         }
      Erosion(IMAGE, MASK, FILTER);
      FILTER->Rows=IMAGE->Rows;
      FILTER->Cols=IMAGE->Cols;
      }
```

SEE ALSO: Binary Dilation, Erosion, and Opening Filters

CLASS: Morphological Filters

DESCRIPTION:

Morphological closing of an image is defined as the graylevel dilation of the image followed by the graylevel erosion of the dilated image. The closing filter operation will reduce small negative oriented graylevel regions and negative graylevel noise regions generated from salt and pepper noise.

$$\text{Close}(A, B) = (A \oplus B) \ominus B$$

EXAMPLES:

(a)

(b)

(a) The original image and ***(b)*** the closed image using an all zero 7×7 mask.

ALGORITHM:

The program assumes that the original image is a 256 graylevel × IMAGE->Rows × IMAGE->Cols pixel image stored in the structure IMAGE. The N × N structuring function is stored in the array MASK[][]. Upon completion of the program, the closed image is stored in the structure FILTER. The graylevel erosion and dilation functions used by the algorithm can be found under graylevel erosion and dilation respectively.

```
#define N 5

CloseGray(struct Image *IMAGE, int
   MASK[][N],struct Image *FILTER)
      {
      int X, Y;
      DilationGray(IMAGE, MASK, FILTER);
      for(Y=0; Y<IMAGE->Rows; Y++)
         {
         for(X=0; X<IMAGE->Cols; X++)
            {
            *(IMAGE->Data+X+(long)Y*
            IMAGE->Cols)= *(FILTER->
            Data+X+(long)Y*IMAGE->Cols);
            }
         }
      ErosionGray(IMAGE, MASK, FILTER);
      FILTER->Rows=IMAGE->Rows;
      FILTER->Cols=IMAGE->Cols;
      }
```

SEE ALSO: Graylevel Dilation, Erosion, Top-Hat, and Opening Filters

CLASS: Mensuration

DESCRIPTION:

Clustering is the process of counting and labeling of objects within an image. Clusters consist of pixel groupings that are related to one another by a predetermined measure. This measure can be defined as a distance between clusters or a similarity measure such as pixel value, or it may be a complex set of identifiers and constraints that define membership in a cluster. The cluster may ultimately be mapped into a binary image as a unique object.

One of the simplest approaches is to specify a size in pixels that the clusters should be. If we assume a binary image, the clustering algorithm processes each pixel and when one is found that is nonzero, it becomes part of the first cluster and is marked. The next nonzero pixel found is tested to see if the distance between it and the previous pixel is less than or equal to the desired cluster pixel size. If it is, it is marked as a member of the first cluster and the search continues. If it is not, it becomes the first member of the second cluster and is marked accordingly. This process continues until all pixels have been evaluated.

The example shows a graphic of a set of similar-sized objects that have been labelled by grayscale value.

EXAMPLE:

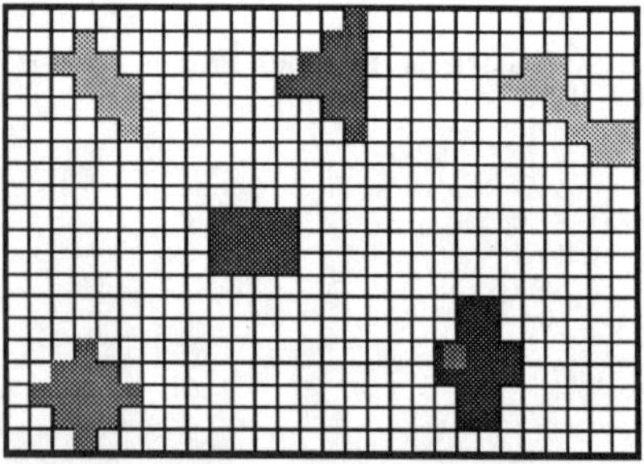

ALGORITHM:

The following variables, with their descriptions, are used in the cluster analysis function:

`i,j,k`	Loop variables.
`n`	Cluster count variable.
`In`	Unsigned character image data structure pointer to *binary* image.
`In->Rows`	Rows in Image
`In->Cols`	Columns in Image
`clustx[]`	Temporary storage for x-coordinate of last pixel in cluster [].
`clusty[]`	See clustx.
`new_clust`	Flag for new cluster.
`edist`	Euclidean distance.
`clust_size`	Size of clusters to be found.

At the completion of the function, the **In** data array contains pixel values that reflect their membership in the clusters found that meet the criteria of being **clust_size** apart. Clusters of pixels that are larger than clust_size size will be partitioned into pieces as they are found in the image. The image is scanned from the upper left corner origin in column-row order.

```
cluster(struct Image *In, int clust_size)
{
   int n, i, j, k, new_clust;
   float edist, clustx[256],clusty[256];

   /* initialize cluster count variable n */
   n=0;

   /* double-loop through each pixel
      of binary image */
   for(i=0; i < In->Rows; ++i)
      for(j=0; j < In->Cols; ++j)
         if(n==0){
```

```
        /* only do for 1st cluster */
           if(pix(In,i,j) !=0){
              n=1;  /* 1st cluster found */

              /* store X coord. */
              clustx[1] = i;
              /* store y coord. */
              clusty[1] = j;

              /* mark pixel as cluster 1 */
              pix(In,i,j) = 1;
           }
        }
        /* test for membership in all
           known clusters */
        else if(pix(In,i,j) != 0){
           /* marker for new cluster */
           new_clust = 0;
           /* compute Euclidean distance */
           for(k = 1; k <= n; ++k){
              edist = sqrt((i-clustx[k])*
                      (i-clustx[k])+
                      (j-clusty[k])*
                      (j-clusty[k]));
              /* test against cluster
                 size desired */
              if(edist <= clust_size){
        /* set pixel to cluster number */
                 pix(In,i,j) =
                      (unsigned char)k;
                 new_clust = 1;
                 k = n + 1;
              }
           }
           /* add a new cluster */
           if(new_clust == 0 && n <= 255){
              n = n + 1;
              clustx[n] = i;
              clusty[n] = j;
              pix(In,i,j) =
                    (unsigned char)n;
           }
        }
}
```

Coding and Compression Class

DESCRIPTION:

Coding is a broad term that can refer to data encryption, representation, or compression schemes. Although images are often encrypted for security purposes, these methods are beyond the scope of this book. Only one representation coding is covered in this book, chain codes. Chain codes play an important role in image understanding and computer vision studies and allow the user to describe contours and edges in a compact and useful way.

Compression is an important aspect of image processing simply because the data sets are so large. Clearly, an RGB, 512 × 512 pixel image at 24 bits per pixel requires 786,432 bytes of storage. Two such pictures exceed the capacity of most magnetic floppy diskettes. If the image data is fairly correlated, the compression ratio possible can exceed 50 to 1. Of course, the basic drawback to compression is the time it takes to compress and the time it takes to decompress. However, special hardware is available that can maximize the speed of the compress/decompress cycle to real-time image display rates.

Although image Storage Formats constitute a separate class in themselves, formats often incorporate compression schemes as a part of their standard.

CLASS MEMBERSHIP:

- Chain Codes
- Huffman Coding
- Run Length Encoding (RLE)

SEE ALSO: Storage Formats

Color Image Processing

DESCRIPTION:

A new and very powerful field of image processing is color image processing. Color image processing can be separated into two general areas: (1) the enhancement of color images, and (2) the pseudocoloring of graylevel images. The enhancement of a color image involves either the enhancement of the three color (Red, Blue, and Green) components of an image or the enhancement of the tint or saturation of an image. In contrast, pseudocoloring makes use of color to represent the graylevel values of an image. In this way, certain attributes of an image can be easily visualized. The human eye is much more sensitive to color variations than to graylevel variations. For example, arteries within an x-ray image can be hilighted in red to improve the accuracy in locating blocked arteries.

CLASS MEMBERSHIP:

- C.I.E. Chromaticity Diagram
- Color Saturation Correction
- Color Tint Correction
- HSI Color Model
- Pseudocolor
- Pseudocolor Display
- RGB Color Model
- True-color Display
- YIQ Color Model

CLASS: Color Image Processing

DESCRIPTION:

This algorithm adjusts the color saturation of an RGB color image in a similar manner as the color control on a color television. The algorithm is based upon the R − Y, G − Y, and B − Y color model used in many of today's color televisions. The color components in terms of R, G, and B are

$$R - Y = 0.70R - 0.59G - 0.11B$$
$$B - Y = -0.30R - 0.59G + 0.89B$$
$$G - Y = -0.30R + 0.41G - 0.11B,$$

where the illuminance component Y is defined as

$$Y = 0.30R + 0.59G + 0.11B$$

ALGORITHM:

The program assumes that the original image is a 24-bit color image of spatial size of RED->Rows × RED->Cols pixels. The program assumes that the image is a RGB image and is stored in three structures: RED, GREEN, and BLUE. Passed to the program in the variable saturation is an integer number that is greater than zero that represents in percent the amount of color to be added back to the illuminance or black and white image. The first thing the program does is compute the R − Y, G − Y, and B − Y components. Next, it scales these values using the variable saturation and adds them back to the illuminance component to generate the new RGB values. Upon completion of the program, the saturated corrected image is returned in the three structures RED, GREEN, and BLUE.

```
Saturation(struct Image *RED, struct Image
      *BLUE, struct Image *GREEN,
      int saturation)
      {
      int X, RY, BY, GY, RY1, BY1, GY1;
      int Y, R, B, G, B1, R1, G1;
      long int K;
```

```
for(Y=0; Y<RED->Rows; Y++)
   {
   for(X=0; X<RED->Cols; X++)
      {
      K=X+(long)Y*RED->Cols;
      R1= *(RED->Data+K);
      G1= *(GREEN->Data+K);
      B1= *(BLUE->Data+K);
      RY1= ((int)(70*(int)R1-
      59*(int)G1-11*(int)B1)/100);
      BY1=((int)(-30*(int)R1-
      59*(int)G1+89*(int)B1)/100);
      GY1=(int)((-30*(int)R1+
      41*(int)G1-11*(int)B1)/100);
      Y=((int)(30*(int)R1+
      59*(int)G1+11*(int)B1)/100);
      BY=(BY1*saturation)/100;
      RY=(RY1*saturation)/100;
      GY=(GY1*saturation)/100;
      R= RY +Y;G= GY + Y;B = BY + Y;
      if(R <0) R=0;
      if(B <0) B=0;
      if(G <0) G=0;
      if(R>255) R=255;
      if(B >255) B=255;
      if(G >255) G=255;
      *(GREEN->Data+K)=
      (unsigned char)G;
      *(BLUE->Data+K)=
      (unsigned char)B;
      *(RED->Data+K)=
      (unsigned char)R;
      }
   }
}
```

SEE ALSO: RGB, HSI and YIQ Color Models, C.I.E. Color Chart, Pseudocolor, and Color Tint Correction

Color Tint Correction

CLASS: Color Image Processing

DESCRIPTION:

This algorithm adjusts the tint of an RGB color image in a similar manner as the tint control on a color television. The algorithm is based upon the R – Y, G – Y, and B – Y color model used in many of today's color televisions. The color components in terms of R, G, and B are

$$R - Y = 0.70R - 0.59G - 0.11B$$
$$B - Y = -0.30R - 0.59G + 0.89B$$
$$G - Y = -0.30R + 0.41G - 0.11B,$$

where the illuminance component Y is defined as

$$Y = 0.30R + 0.59G + 0.11B$$

ALGORITHM:

The program assumes that the original image is a 24-bit color image of spatial size of RED->Rows × RED->Cols pixels. The program assumes that the image is a RGB image and is stored in three equal sized structures: RED, GREEN, and BLUE. Passed to the program in the variable tint is an integer number in degrees between -180 and +180 that represents the angle in which the tint of the image is to be rotated. Positive angles give the color image more of a green tint while negative angles give the color image more of a red tint. The first thing the program does is compute the R – Y and B – Y components. Next, it rotates these two vectors by the angle given in the variable tint. Finally, the program then recomputes the new RGB values placing them in the three color image arrays. Upon completion of the program, the tint corrected image is returned in the three structures RED, GREEN, and BLUE.

```
Hue(struct Image *RED, struct Image  *BLUE,
struct Image *GREEN, int  tint)
     {
     int X, RY, BY, GY, RY1, BY1;
     int Y, R, B, G, S, C, R1, G1, B1;
     long int K;
```

```
float theta;
theta=(3.14159*(float)tint)/180.0;
C=256*cos((double)theta);
S=256*sin((double)theta);
for(Y=0; Y<RED->Rows; Y++)
   {
   for(X=0; X<RED->Cols; X++)
      {
      K=X+(long)Y*RED->Cols;
      R1= *(RED->Data+K);
      G1= *(GREEN->Data+K);
      B1= *(BLUE->Data+K);
      RY1= ((int)(70*(int)R1-
      59*(int)G1-11*(int)B1)/100);
      BY1=((int)(-30*(int)R1-
      59*(int)G1+89*(int)B1)/100);
      Y=((int)( 30*(int)R1+
      59*(int)G1+11*(int)B1)/100);
      BY=(C*BY1 - S*RY1)/256;
      RY=(S*BY1 + C*RY1)/256;
      GY=((int)(-51*RY -
      19*BY)/100);
      R= RY +Y;G= GY + Y;B = BY + Y;
      if(R <0) R=0;
      if(B <0) B=0;
      if(G <0) G=0;
      if(R>255) R=255;
      if(B >255) B=255;
      if(G >255) G=255;
      *(GREEN->Data+K)=
      (unsigned char)G;
      *(BLUE->Data+K)=
      (unsigned char)B;
      *(RED->Data+K)=
      (unsigned char)R;
      }
   }
}
```

SEE ALSO: RGB, HSI and YIQ Color Models, C.I.E. Color Chart, Pseudocolor, and Color Saturation Correction

CLASS: Mensuration

DESCRIPTION:

Compactness is an object distribution measure similar to circularity, but is defined as the perimeter squared divided by the area. The compactness value will be minimal for a circular object, but it is not insensitive to noise in the perimeter measure. Objects with noisy boundaries will have a large perimeter measure with respect to similar sized objects with smooth boundaries, hence different compactness values in spite of circular shape. No algorithm is given as compactness can be readily calculated from the area of the object and the area of the object boundary (perimeter). Examples are shown, however, of compactness measurements on various objects.

EXAMPLE:

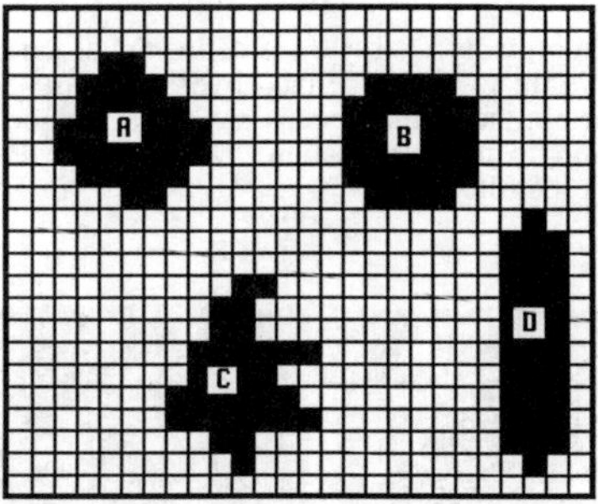

Each of the objects shown has the same area, 32 pixels. Compactness for each object is as follows:

A: 7.03 B: 8 C: 15.15 D: 15.15

Objects A and B are similarly shaped and have very close compactness values. Objects C and D, however, are very different in shape in spite of identical compactness values.

SEE ALSO: Area, Circularity, Perimeter

Contra-Harmonic Mean Filter

CLASS: Nonlinear Filters

DESCRIPTION:

The contra-harmonic mean filter is member of a set of nonlinear mean filters which are better at removing Gaussian type noise and preserving edge features than the arithmetic mean filter. The contra-harmonic mean filter is very good at removing positive outliers for negative values of P and negative outliers for positive values of P. If all the pixels included in the calculation of the contra-harmonic mean are zero, the output of the contra-harmonic filter will also be zero. The definition of the contra-harmonic mean filter is

$$\text{contra-harmonic Mean}(A) = \frac{\sum_{(i,j) \in M} A(x+i,\ y+j)^{P+1}}{\sum_{(i,j) \in M} A(x+i,\ y+j)^{P}},$$

where the coordinate x + i, y + j is defined over the image A and the coordinate i, j is defined over the mask M. The mask M determines which pixels are to be included in the contra-harmonic mean calculation. The parameter P chooses the order of the filter.

EXAMPLE:

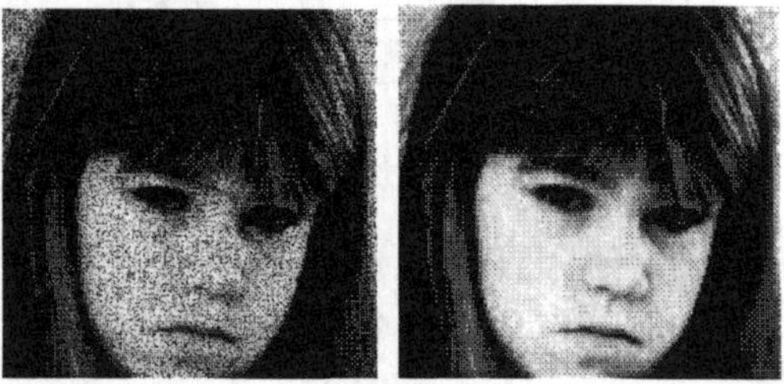

(a) *(b)*

(a) The 10% negative outlier corrupted image and *(b)* the contra-harmonic mean filtered image (3 × 3 mask, P = 2).

ALGORITHM:

The program assumes that the original image is a 256 graylevel × IMAGE->Rows × IMAGE->Cols pixel image stored in the structure IMAGE. The program computes the contra-harmonic filter over a set of pixels contained within a square N × N region of the image centered at the pixel X, Y. The program expects the order of the filter to be passed to it upon execution. The size of the filtering operation is determined by the variable N and should be set to an odd number and should be less than 12. Upon completion of the program, the filtered image is stored in the structure IMAGE1.

```
ContraharmonicMean(struct Image *IMAGE,
   int P, struct Image *IMAGE1)
      {
      int X, Y, I, J, Z;
      int N;
      int AR[121], A;
      float SUM;
      float SUM1;
      N=5;
      for(Y=N/2; Y<IMAGE->Rows-N/2; Y++)
         for(X=N/2; X<=IMAGE->Cols-N/2; X++)
            {
            Z=0;
            for(J=-N/2; J<=N/2; J++)
               for(I=-N/2; I<=N/2; I++)
                  {
                  AR[Z]=*(IMAGE->Data+X
                  +I+(long)(Y+J)
                  *IMAGE->Cols);
                  Z++;
                  }
            Z=0;
            SUM=0.0;
            SUM1=0.0;
            for(J=0; J<=N*N-1;J++)
               {
               if(AR[J]==0 && P<0)
                  Z=1;
               else
                  {
                  SUM=SUM+pow((double)AR[J],
                  (double)(P+1));
```

```
                SUM1=SUM1+pow((double)
                AR[J], (double)P);
                }
            }
        if(Z==1)
           *(IMAGE1->Data+X +(long)Y
           *IMAGE->Cols)=0;
        else
           {
           if(SUM1==0.0)
             A=0.0;
           else
             A=(int)(SUM/SUM1);
           if(A >255)
                A = 255;
           *(IMAGE1->Data+X+(long)Y
           *IMAGE->Cols) =
           (unsigned char)A;
          }
        }
    }
```

SEE ALSO: Geometric, Y_p, Harmonic and Arithmetic mean Filters, Median, and other Nonlinear Filters

CLASS: Histogram Operation

DESCRIPTION:

Contrast correction is used to enhance the visual appearance of an image. Contrast modification of an image is defined as

$$s_i = \text{contrast} \cdot (g_i - \text{average}) + \text{average} \, ,$$

where s_i is the ith graylevel value of the contrast enhanced image, g_i is the ith graylevel value of the original image, and average is the mean value of the original image given by

$$\text{average} = \frac{1}{N_x N_y} \sum_{X=0}^{N_x} \sum_{Y=0}^{N_y} P(X.\ Y) \ ,$$

where N_x and N_y are the dimension of the image in the x and y directions respectively, and P(X , Y) is the graylevel value of the pixel at the X, Y coordinate.

EXAMPLE:

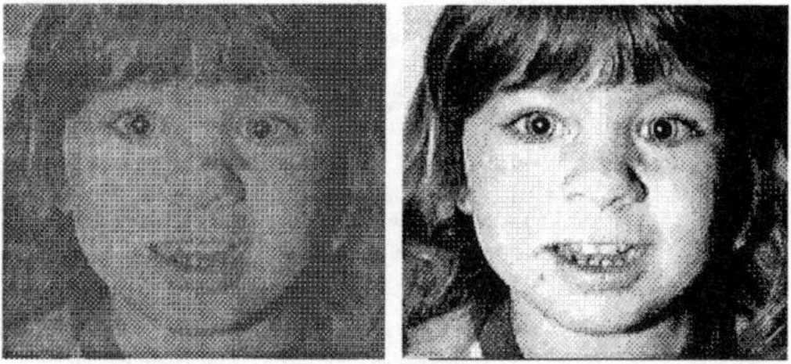

(a) *(b)*

(a) The original image and ***(b)*** the contrast corrected image.

ALGORITHM:

The program assumes that the original image is a 256 graylevel × IMAGE->Rows × IMAGE->Cols pixel image stored in the structure IMAGE. The program then computes the average of the image and removes this value from the original pixel. Next, the averaged removed pixel is multiplied by the specified contrast parameter. The last step the program implements is to add back the average to the contrast corrected pixel and store this result in the structure IMAGE1.

```
Contrast(struct Image *IMAGE, struct Image
*IMAGE1, float contrast)
      {
      int X, Y, I;
      long int SUM, J, R;
      float AVG;
      J=0;
      R=IMAGE->Cols;
      SUM=0;
      for(Y=0; Y<=IMAGE->Rows; Y++)
         for(X=0; X<=IMAGE->Cols; X++)
            {
            SUM=SUM +
            *(IMAGE->Data+X+(long)Y*R);
            J++;
            }
      AVG=(float)SUM/(float)J;
      for(Y=0; Y<IMAGE->Rows; Y++)
         {
         for(X=0; X<IMAGE->Cols; X++)
            {
            I= contrast*((float)
            *(IMAGE->Data+X+(long)Y*R)-
            AVG)+AVG;
            if(I> 255)
              I=255;
            if(I<0)
              I=0;
            *(IMAGE1->Data+X+(long)Y*R) = I;
            }
         }
      }
```

SEE ALSO: Graylevel Histogram, Histogram Specification, Brightness Correction and Nonlinear Graylevel Transformations

Dilation (Binary) Filter

CLASS: Morphological Filters

DESCRIPTION:

Derived from Minkowski addition, binary dilation of an object increases its geometrical area by setting the background pixels adjacent to an object's contour to the object's graylevel value. Dilation is defined as the union of all vector additions of all pixels *a* in object *A* with all pixels *b* in the structuring function *B*.

$$A \oplus B = \{t \in Z^2 : t = a + b, a \in A, b \in B\},$$

where the vector t is an element of the image space Z^2.

EXAMPLES:

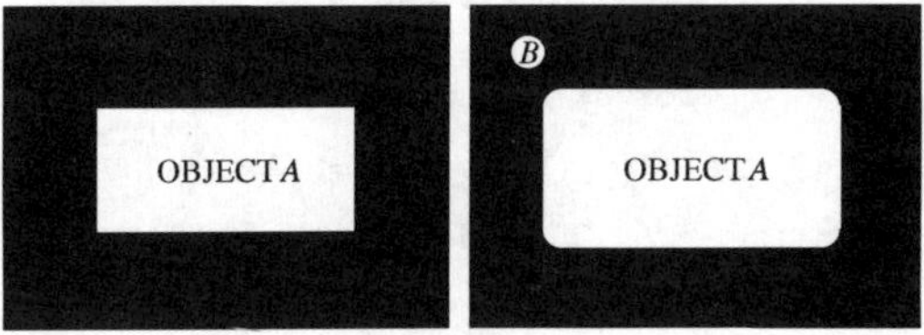

(a) *(b)*

(a) The binary image of a rectangle and *(b)* the dilated image of object *A* with a circular structuring function *B*.

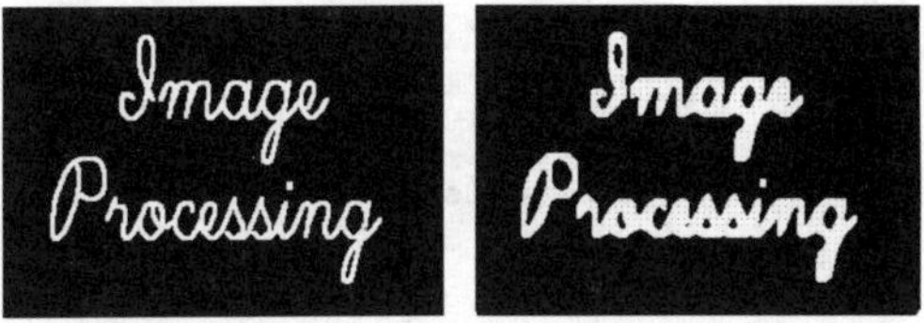

(a) *(b)*

(a) The original image and *(b)* the dilated image.

ALGORITHM:

The program assumes that the original binary image is an IMAGE->Rows × IMAGE->Cols pixel image with a background graylevel value of 0 and an object graylevel value of 255 (object) and is stored in the structure IMAGE. The N × N structuring function is stored in array MASK[][]. Upon completion of the program, the dilated image is stored in the structure FILTER.

#define N 5

```
Dilation(struct Image *IMAGE, int
   MASK[][N], struct Image *FILTER)
      {
      int X, Y, I, J, smax;
      for(Y=N/2; Y<IMAGE->Rows-N/2; Y++)
         for(X=N/2; X<IMAGE->Cols-N/2; X++)
            {
            smax=0;
            for(J=-N/2; J<=N/2; J++)
               {
               for(I=-N/2; I<=N/2; I++)
                  {
                  if(MASK[I+N/2][J+N/2] ==1)
                    {
                    if(*(IMAGE->Data+X+I+
                    (long)(Y+J)
                    *IMAGE->Cols)>smax)
                      {
                     smax=*(IMAGE->Data+X+I+
                     (long)(Y+J)*IMAGE->Cols);
                      }
                    }
                  }
               }
            *(FILTER->Data+ X +
            (long)Y * IMAGE->Cols) = smax;
            FILTER->Rows=IMAGE->Rows;
            FILTER->Cols=IMAGE->Cols;
            }
      }
```

SEE ALSO: Maximum and Minimum Filters, Binary, Erosion, Opening, and Closing Filters

Dilation (Graylevel) Filter

CLASS: Morphological Filters

DESCRIPTION:

Graylevel dilation of an image is used to smooth small negative graylevel regions. Graylevel dilation is defined as the maximum of the sum of a local region of an image and a given graylevel mask

$$A \oplus B = \max[\, A(\mathrm{x} + \mathrm{i}, \mathrm{y} + \mathrm{j}) + B(\mathrm{i}, \mathrm{j}) \,] ,$$

where the coordinate x + i, y + j is defined over the image A and the coordinate i, j is defined over the mask B. For the special case when all the mask values are 0, graylevel dilation reduces to the maximum filter.

EXAMPLES:

(a)

(b)

(a) The original image and ***(b)*** the dilated image using an all zero 3 × 3 mask applied recursively 3 times.

ALGORITHM:

The program assumes that the original image is a 256 graylevel × IMAGE->Rows × IMAGE->Cols pixel image stored in the structure IMAGE. The N × N structuring function is stored in the array MASK[][]. Upon completion of the program, the dilated image is stored in the structure FILTER.

#define N 5

```
DilationGray(struct Image *IMAGE,int
   MASK[][N], struct Image *FILTER)
      {
      int a[N][N];
      int X, Y, I, J, smax;
      for(Y=N/2; Y<IMAGE->Rows-N/2; Y++)
         for(X=N/2; X<IMAGE->Cols-N/2; X++)
            {
            smax=0;
            for(J=-N/2; J<=N/2; J++)
              for(I=-N/2; I<=N/2; I++)
                 a[I+N/2][J+N/2]=
                 *(IMAGE->Data
                 +X+I+(long)(Y+J)
                 *IMAGE->Cols)+
                 MASK[I+N/2][J+N/2];
            for(J=-N/2; J<=N/2; J++)
               {
               for(I=-N/2; I<=N/2; I++)
                  {
                  if(a[I+N/2][J+N/2] > smax)
                    smax = a[I+N/2][J+N/2];
                  }
               }
            if(smax>255)
                 smax=255;
            *(FILTER->Data + X+(long)Y*
            IMAGE->Cols) =smax;
            }
      FILTER->Rows=IMAGE->Rows;
      FILTER->Cols=IMAGE->Cols;
      }
```

SEE ALSO: Graylevel Erosion, Opening, Closing and Top-Hat, and Minimum Filters

Discrete Convolution

CLASS: Image Fundamentals

DESCRIPTION:

Discrete Convolution is best described as a combining process that copies one image into another and is the fundamental result of a filtering operation. Any number of filters may be applied to an image by convolving the filter *mask* with the original image. The equation for the two-dimensional discrete convolution is given by:

$$\mathrm{Out}(i,j) = \sum_{m=0}^{M-1} \sum_{n=0}^{N-1} \mathrm{In}(m,n)\mathrm{Mask}(i-m,j-n)$$

Where **In** is the input image, **Mask** is the convolution mask, and **Out** is the output image. The dimension of the images is **M** by **N**, the **Mask** image is normally of substantially smaller size than **M** by **N**; however, the mask is filled in with zeroes to allow for consistency in the indexing.

The convolution algorithm will generate results that are greater than the range of original values of the input image. Therefore, a *scaling* operation is required to restore the result to the same graylevel range of the original picture.

The example below shows the original image (256 × 256) on the left and the result of convolution with a 5 × 5 sharpening mask on the right. Note the border around the convolved result; this is a side effect of the algorithm as the overlaying mask passes over the boundary of the original image.

EXAMPLE:

ALGORITHM:

The algorithm computes the two-dimensional discrete convolution between two images, **In** and **Mask**, and leaves the result in the image **Out**. It is assumed that the dimensions of the images are congruent and that they are of unsigned character type. Note that overflow and underflow is truncated. This routine is easily modified to operate on integer or real images followed by a scaling procedure.

```
/* 2-D Discrete Convolution */
void Convolve(struct Image *In,
              struct Image *Mask,
              struct Image *Out)
{
   long i,j,m,n,idx,jdx;
   int ms,im,val;
   unsigned char *tmp;

   for(i=0;i<In->Rows;++i)
      for(j=0;j<In->Cols;++j){
         val = 0;
         for(m=0;m<Mask->Rows;++m)
            for(n=0;n<Mask->Cols;++n){
               ms = (signed char)*(Mask->Data
                     + m*Mask->Rows + n);
               idx = i-m;
               jdx = j-n;
               if(idx>=0 && jdx>=0)
                  im = *(In->Data +
                     idx*In->Rows + jdx);
               val += ms*im;
            }
         if(val > 255)val = 255;
         if(val < 0)  val = 0;
         tmp = Out->Data + i*Out->Rows + j;
         *tmp = (unsigned char)val;

      }
}
```

SEE ALSO: Discrete Correlation, Mask, Scaling, Spatial Filters

CLASS: Image Fundamentals

DESCRIPTION:

Discrete Correlation is a process by which one image is compared mathematically with another. The resulting image is a two-dimensional expression of equivalence. The equation for the two-dimensional discrete correlation is given by:

$$Out(i,j) = \sum_{m=0}^{M-1} \sum_{n=0}^{N-1} In1(m,n)In2(i+m,j+n)$$

Where **In1** and **In2** are the input images and **Out** is the output image. The dimension of the images is **M** by **N**, the input images may be of different dimensions; however, the sizes are filled in with zeroes so that the images going into the algorithm are M × N to allow for consistency in the indexing.

The second image, **In2**, is often called the *template* and the correlation is then called *template* matching. The output picture's maximum value will reveal the spatial postion of greatest match between the original image and the template, as illustrated in the example below, where the dotted line surrounding **UCF** in the first picture is the match template. The second picture has a maximum at the location of best match, or correlation between the template and original image. Since the result of correlation produces values that may be out of the range of the original picture grayscale, *scaling* should be applied to the result.

EXAMPLE:

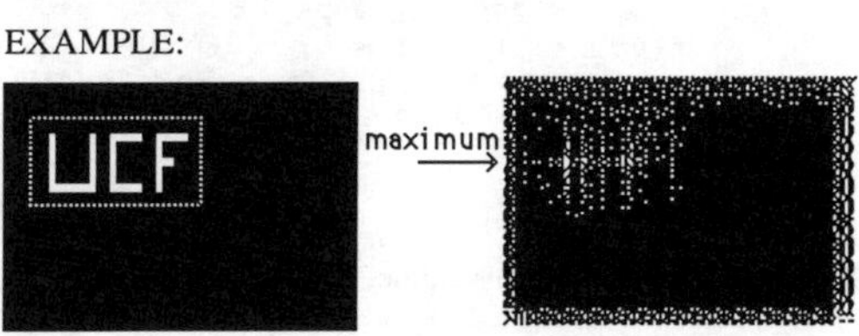

ALGORITHM:

The algorithm computes the two-dimensional discrete correlation between an image, **In** and a template **Tmpl**, and leaves the result in the image **Out**. It is assumed that the dimensions of the images are congruent and that they are of type unsigned character. Note that overflow and underflow is truncated. This routine is easily modified to operate on integer or real images followed by a scaling procedure.

```
/* 2-D Discrete Correlation */
void Correlate(struct Image *In,
               struct Image *Tmpl,
               struct Image *Out)
{
   long i,j,m,n,idx,jdx;
   int tm,im,val;
   unsigned char *tmp;

   /* the outer summation loop */
   for(i=0;i<In->Rows;++i)
      for(j=0;j<In->Cols;++j){
         val = 0;
         for(m=0;m<Tmpl->Rows;++m)
            for(n=0;n<Tmpl->Cols;++n){
               tm = (signed char)*(Tmpl->Data
                     + m*Tmpl->Rows + n);
               idx = i+m;
               jdx = j+n;
               if(idx>=0 && jdx>=0)
                  im = *(In->Data +
                        idx*In->Rows + jdx);
               val += tm*im;
            }
         if(val > 255)val = 255;
         if(val < 0)  val = 0;
         tmp = Out->Data + i*Out->Rows + j;
         *tmp = (unsigned char)val;

   }
}
```

SEE ALSO: Discrete Convolution, Scaling

CLASS: Transform

DESCRIPTION:

The discrete cosine transform is used in image coding and compression. One nice feature of the discrete cosine transform is that it can be easily computed in the same manner as the two-dimensional Fourier transform. This transform is very similar to the real part of the Fourier transform.

The discrete cosine transform defined in two dimensions is given by

$$C(n, m) = k(n)\,k(m) \sum_{Y=0}^{N-1} \sum_{X=0}^{N-1} f(X, Y) \cdot$$

$$\cos\left(\pi n \frac{2X+1}{2N}\right) \cdot \cos\left(\pi m \frac{2Y+1}{2N}\right)$$

where n and m are defined from 1 to N − 1, and

$$k(n) = \begin{cases} \sqrt{\frac{1}{N}} & \text{for } n = 0 \\ \sqrt{\frac{2}{N}} & \text{otherwise} \end{cases}$$

The inverse discrete cosine transform is given by

$$f(X,Y) = \sum_{n=0}^{N-1} \sum_{m=0}^{N-1} k(n) \cdot k(m) \cdot F(n, m) \cdot$$

$$\cos\left(\pi n \frac{2X+1}{2N}\right) \cdot \cos\left(\pi m \frac{2Y+1}{2N}\right)$$

ALGORITHM:

The program assumes that the original image is a floating point <u>square</u> image of size IMAGE->Rows stored in the structure IMAGE. Passed to the program is the *dir* variable. This variable determines if a forward (dir = 1) or if an inverse (dir = -1) discrete cosine transform is to be performed. The program first computes the discrete cosine transform in the x direction followed by the discrete cosine transform in the y direction. Upon completion of the program, the discrete cosine components are returned in the floating point structure IMAGE1.

```
DiscreteCosine(struct  Image  *IMAGE,struct
Image *IMAGE1,  int dir)
      {
      int X, Y, n, m, num;
      float sum, pi,k0,k1, ktx, kty, A;
      pi=3.141592654;
      num=IMAGE->Rows;
      k0=sqrt((double)1.0/(double)num);
      k1=sqrt((double)2.0/(double)num);
      for(m=0; m<num; m++)
         for(n=0; n<num; n++)
            {
            sum=0.0;
            for(Y=0; Y<num; Y++)
               {
               if(dir==1)
                    A=cos((double)((2.0*
                    (float)Y+1)*
                    m*pi/2.0/num));
               if(dir==-1)
                    A=cos((double)((2.0*
                    (float)m+1)*
                    Y*pi/2.0/num));
               for(X=0; X<num; X++)
                  {
                  if(dir==-1)
                    {
                    if(X==0)
                      ktx=k0;
                    else
                      ktx=k1;
                    if(Y==0)
                      kty=k0;
                    else
```

```
                        kty=k1;
                        sum=sum + *(IMAGE->Data +
                        X+(long)Y* IMAGE->Rows)*
                        cos((double)((2.0*(float)
                        n+1)*X*pi/2.0/(float)num)
                        )*A*ktx*kty;
                        }
                     if(dir==1)
                        {
                        ktx=1;
                        kty=1;
                        sum=sum + *(IMAGE->Data +
                        X+(long)Y* IMAGE->Rows)*
                        cos((double)((2.0*(float)
                        X+1)*n*pi/2.0/(float)num)
                        )*A*ktx*kty;
                        }
                     }
                  }
               if(dir==1)
                  {
                  if(n==0)
                     sum=sum*k0;
                  else
                     sum=sum*k1;
                  if(m==0)
                     sum=sum*k0;
                  else
                     sum=sum*k1;
                  }
               *(IMAGE1->Data + n + (long)m *
               IMAGE->Rows) = sum;
               }
               IMAGE1->Rows=IMAGE->Rows;
               IMAGE1->Cols=IMAGE->Cols;
         }
```

SEE ALSO: Fourier Transform Properties, Hadamard and Walsh Transforms, and the Discrete Fourier Transform

CLASS: Transform

DESCRIPTION:

The mathematical formulation of a function as a series of sine and cosine functions was first developed by the French mathematician Baptiste Joseph Fourier (1768-1830). The discrete Fourier transform decomposes an image into a set of cosines and sines each of which represents the frequency components of an image. Assuming an image of size X = N by Y = M pixels, the two-dimensional and its inverse discrete Fourier transform are defined as

$$F(n, m) = \frac{1}{NM} \sum_{Y=0}^{N-1} \sum_{X=0}^{N-1} f(X, Y) \cdot$$

$$[\cos(2\pi nX/N) + j \sin(2\pi nX/N)] \cdot$$

$$[\cos(2\pi mY/M) + j \sin(2\pi mY/M)]$$

$$f(X, Y) = \sum_{n=0}^{N-1} \sum_{m=0}^{N-1} F(n, m) \cdot$$

$$[\cos(2\pi nX/N) - j \sin(2\pi nX/N)] \cdot$$

$$[\cos(2\pi mY/M) - j \sin(2\pi mY/M)],$$

where both f(x, y) and F(n, m) are complex two-dimensional functions.

Given the real and imaginary Fourier frequency components of F(n, m), the magnitude and phase terms are defined as

$$|F(n, m)| = \sqrt{\{F_r(n, m)\}^2 + \{F_i(n, m)\}^2}$$

$$\text{ang}[F(n, m)] = \arctan\left[\frac{F_i(n, m)}{F_r(n, m)}\right]$$

The function |F(n, m)| is known as the magnitude spectrum of f(x, y), and the function ang[F(n, m)] is known as the phase spectrum of f(x, y).

EXAMPLE:

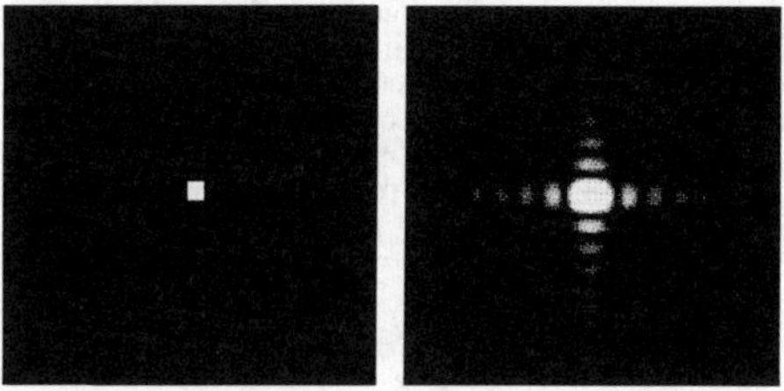

(a) The original image of a square and ***(b)*** the two-dimensional DFT magnitude image.

ALGORITHM:

The program assumes that the original image is a complex floating point square image of size IMAGE->Rows stored in the structure IMAGE. The maximum allowable size of the discrete Fourier transform is limited to a 512 × 512 transform. Adjacent elements in the structure IMAGE are the real and imaginary components of the image F(X, Y) as shown below.

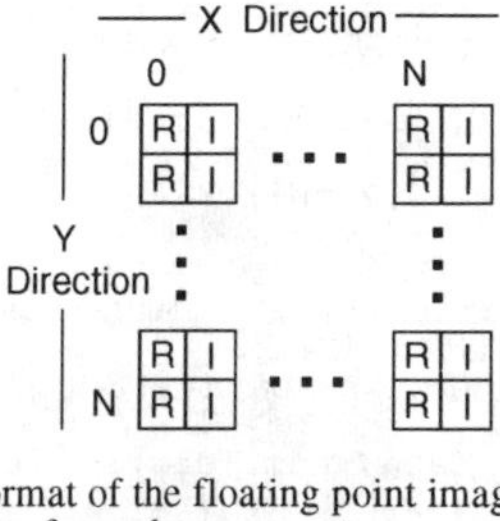

Storage format of the floating point image and its Fourier transformed components.

For example, the first data value given is the real component while the second data value given is the imaginary component of the first pixel of the image. Passed to the program is the *dir* variable. This variable determines if a forward (dir = 1) or if an inverse (dir = -1) Fourier transform is to be performed. The program first computes the discrete Fourier transform in the x direction followed by the discrete Fourier transform in the y direction using the Fast Fourier Transform algorithm. Upon completion of the program, the real and imaginary Fourier components are returned in the structure IMAGE. Every other adjacent element in the structure IMAGE represents the real and imaginary Fourier frequency components of F(n, m).

```
DiscreteFourier(struct Image *IMAGE,
      float dir)
      {
      int X, Y, num;
      long int R;
      float data[1024], scale;
      num=IMAGE->Rows;
      if(dir == 1.0)
        scale= num*num;
      else
        scale=1.0;
      for(Y=0; Y<num; Y++)
         {
         R=(long)Y*IMAGE->Rows*2;
         for(X=0;X<=2*num-1; X++)
            data[X]=*(IMAGE->Data+X+R);
         fft(data, num, dir);
         for(X=0;X<=2*num-1; X++)
            *(IMAGE->Data+X+R)=data[X];
         }
      for(X=0; X<=2*num-1; X+=2)
         {
         for(Y=0;Y<num; Y++)
            {
            R=(long)Y*IMAGE->Rows*2;
            data[2*Y]= *(IMAGE->Data+X+R);
            data[2*Y+1]=
            *(IMAGE->Data+X+1+R);
            }
         fft(data, num, dir);
         for(Y=0;Y<num; Y++)
            {
```

```
            R=(long)Y*IMAGE->Rows*2;
            *(IMAGE->Data+X+R)=data[2*Y]/
            scale;
            *(IMAGE->Data+X+1+R)
            =data[2*Y+1]/scale;
            }
        }
    }

fft(float data[],int num, float dir)
    {
    int array_size, bits, ind, j, j1;
    int i, i1, u, step, inc;
    float sine[513], cose[513];
    float wcos, wsin, tr, ti, temp;
    bits=log(num)/log(2)+.5;
    for(i=0;i<num+1;i++)
     {
     sine[i]=dir*sin(3.141592654*i/num);
     cose[i]=cos(3.141592654*i/num);
     }
    array_size=num<<1;
    for(i=0;i<num;i++)
       {
       ind=0;
       for(j=0;j<bits;j++)
          {
          u=1<<j;ind=(ind<<1)+((u&i)>>j);
          }
          ind=ind<<1;j=i<<1;
       if(j<ind)
        {
        temp=data[j]; data[j]=data[ind];
        data[ind]=temp;temp=data[j+1];
        data[j+1]=data[ind+1];
        data[ind+1]=temp;
        }
       }
    for(inc=2;inc<array_size;inc=step)
       {
       step=inc<<1;
       for(u=0;u<inc;u+=2)
          {
          ind=((long)u<<bits)/inc;
          wcos=cose[ind];wsin=sine[ind];
          for(i=u;i<array_size;i+=step)
             {
```

```
            j=i+inc; j1=j+1; i1=i+1;
            tr=wcos*data[j]-
              wsin*data[j1];
            ti=wcos*data[j1]+
              wsin*data[j];
            data[j]=data[i]-tr;
            data[i]=data[i]+tr;
            data[j1]=data[i1]-ti;
            data[i1]=data[i1]+ti;
            }
        }
    }
}
```

SEE ALSO: Fourier Transform Properties, Hadamard and Walsh Transforms, and the Discrete Cosine Transform

Dithering

CLASS: Graphics

DESCRIPTION:

Dithering is the process of producing graylevel images on printers capable of only printing black dots of one intensity on white paper. The process uses a pattern of dots to represent a given graylevel value. The higher the black dot concentration in a fixed area the darker the area will appear. Shown below are 33 graylevels represented by 33 unique dot patterns.

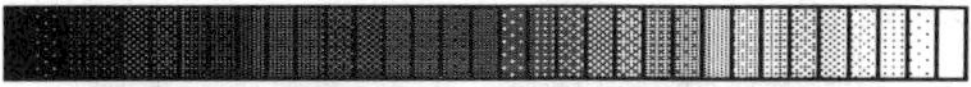

EXAMPLE:

(a)

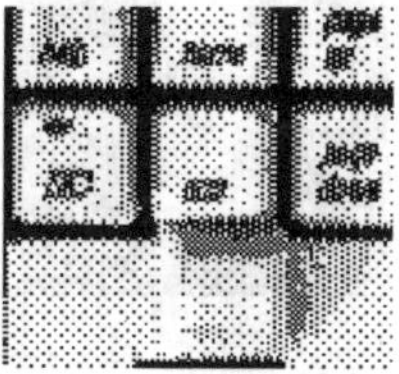

(b)

(c)

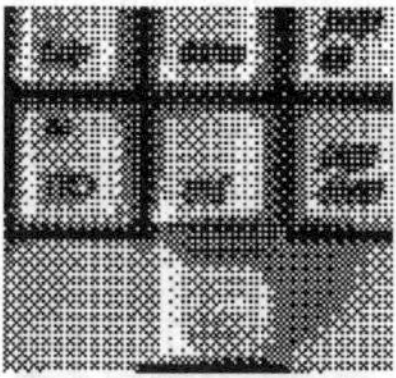

(d)

An example of 3 different dithering patterns: ***(a)*** the original graylevel image and ***(b) - (d)*** the three dithering patterns.

SEE ALSO: Morphing, Warping, and Zooming

CLASS: Morphological Filters

DESCRIPTION:

Derived from Minkowski subtraction, binary erosion of an object reduces its geometrical area by setting the contour pixels of an object to the background value. Erosion is defined as the complement of the resulting dilation of the compliment of object A with structuring function B.

$$A \ominus B = (A^c \oplus B)^c$$

EXAMPLES:

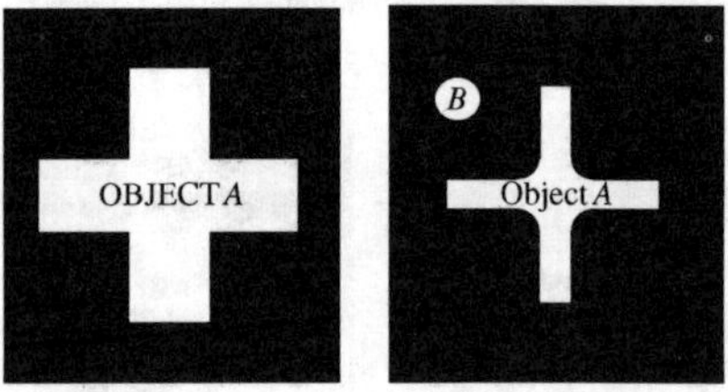

(a) ***(b)***

(a) The original binary image of a cross and ***(b)*** the eroded image of object A with a circular structuring function B.

(a) ***(b)***

(a) The original image and ***(b)*** the eroded image.

ALGORITHM:

The program assumes that the original binary image is an IMAGE->Rows × IMAGE->Cols pixel image with a background graylevel value of 0 and an object graylevel value of 255 (object) and is stored in the structure IMAGE. The N × N structuring function is stored in the array MASK[][]. Upon completion of the program, the eroded image is stored in the structure FILTER.

#define N 5

```
Erosion(struct Image *IMAGE, int MASK[][N],
   struct Image *FILTER)
      {
      int X, Y, I, J, smin;
      for(Y=N/2; Y<IMAGE->Rows-N/2; Y++)
         for(X=N/2; X<IMAGE->Cols-N/2; X++)
            {
            smin=255;
            for(J=-N/2; J<=N/2; J++)
               {
               for(I=-N/2; I<=N/2; I++)
                  {
                  if(MASK[I+N/2][J+N/2]==1)
                     {
                     if(*(IMAGE->Data+X+I+
                     (long)(Y+J)*IMAGE->Cols)
                     < smin)
                       {
                      smin= *(IMAGE->Data+X+I+
                      (long)(Y+J)*IMAGE->Cols);
                       }
                     }
                  }
               }
            *(FILTER->Data+ X +
            (long)Y * IMAGE->Cols) = smin;
            FILTER->Rows=IMAGE->Rows;
            FILTER->Cols=IMAGE->Cols;
            }
      }
```

SEE ALSO: Minimum and Maximum Filters, Binary Dilation, Opening, and Closing Filters

CLASS: Morphological Filters

DESCRIPTION:

Graylevel erosion of an image is used to smooth small positive graylevel regions. Graylevel erosion is defined as the minimum of the difference between a local region of an image and a given graylevel mask

$$A \ominus B = \min[\, A(x+i,\, y+j) - B(i,j)\,]\,,$$

where the coordinate x + i, y + j is defined over the image A and the coordinate i, j is defined over the mask B. For the special case when all the mask values are 0, graylevel erosion reduces to the minimum filter.

EXAMPLES:

(a)

(b)

(a) The original image and ***(b)*** the eroded image using an all zero 3×3 mask applied recursively 3 times.

ALGORITHM:

The program assumes that the original image is a 256 graylevel × IMAGE->Rows × IMAGE->Cols pixel image stored in the structure IMAGE. The N × N structuring function is stored in the array MASK[][]. Upon completion of the program, the eroded image is stored in the structure FILTER.

#define N 5

```
ErosionGray(struct Image *IMAGE,int
   MASK[][N], struct Image *FILTER)
      {
      int a[N][N];
      int X, Y, I, J, smin;
      for(Y=N/2; Y<IMAGE->Rows-N/2; Y++)
         for(X=N/2; X<IMAGE->Cols-N/2; X++)
            {
            smin=255;
            for(J=-N/2; J<=N/2; J++)
              for(I=-N/2; I<=N/2; I++)
                 a[I+N/2][J+N/2]=
                 *(IMAGE->Data
                 +X+I+(long)(Y+J)
                 *IMAGE->Cols)-
                 MASK[I+N/2][J+N/2];
            for(J=-N/2; J<=N/2; J++)
               {
               for(I=-N/2; I<=N/2; I++)
                  {
                  if(a[I+N/2][J+N/2] < smin)
                    smin = a[I+N/2][J+N/2];
                  }
               }
            if(smin<0)
                 smin=0;
            *(FILTER->Data + X+(long)Y*
            IMAGE->Cols) =smin;
            }
      FILTER->Rows=IMAGE->Rows;
      FILTER->Cols=IMAGE->Cols;
      }
```

SEE ALSO: Graylevel Dilation, Opening, Closing and Top-Hat, and Minimum Filters

CLASS: Graphics Algorithms

DESCRIPTION:

Flipping is different from rotation, a similar operation, in that the image is mirrored to itself. The example shows a top-to-bottom flip; notice the word *Panic* is now backwards.

EXAMPLE:

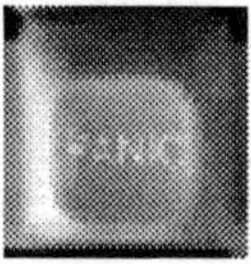

ALGORITHM:

The routine **flip** flips the image passed by the image data structure pointer **In** vertically into the image data structure **Out**. The macro **idx** is used for simplified access to the pixels of the image by coordinates.

```
#define idx(Im,i,j) \
        *(Im->Data + (i)*Im->Cols + (j))

flip(struct Image *In, struct Image *Out)
{
   int i,j,k;

   k  = In->Rows;
   for(i=0; i < In->Rows; ++i){
      for(j=0; j< In->Cols; ++j)
         idx(Out,k,j) = idx(In,i,j);
      --k;
   }
}
```

SEE ALSO: Rotate

CLASS: Transforms

DESCRIPTION: Several important properties of discrete Fourier transforms permit the manipulation of the frequency components of an image. Listed here is a summary of several important discrete Fourier transform properties.

Let f(x, y) and g(x, y) be the original graylevel images of size $N \times M$, and F(n, m) and G(n, m) be their discrete Fourier transform components, respectively.

f(X, Y) IMAGE	F(n, m) MAGNITUDE
$A \cos 2\pi aX \cdot \cos 2\pi bY$	A/4 at (a, b) A/4 at (−a, b) A/4 at (a, −b) A/4 at (−a, −b)
A	A at n = 0, m = 0
A at X = 0, Y = 0 0 elsewhere	$\frac{A}{NM}$
$f(X, Y)(-1)^{(X + Y)}$	F(n, m) centered
f(aX, bY)	$1/\lvert a\, b\rvert\, F(n / a, m / b)$
f(X, Y) + g(X, Y)	F(n, m) + G(n, m)
$\exp[-\pi(X^2 + Y^2)]$	$\exp[-\pi(n^2 + m^2)/N]$
F(n, m)	f(−X,− Y)
f(X, Y) rotated by an angle θ	F(n, m) rotated by an angle θ

SEE ALSO: Discrete Fourier Transforms, Hadamard and Walsh Transforms, and the Discrete Cosine Transform

CLASS: Noise

DESCRIPTION:

Gamma type noise is the result of lowpass filtering an image containing negative exponential noise as the result of acquiring an image which is illuminated by a coherent laser. Its histogram is defined as

$$h_i = \frac{G_i^{\alpha - 1}}{(\alpha - 1)!\, a^{\alpha}} \exp^{-(G_i/a)} \qquad \text{for } 0 \le G_i < \infty,\ \alpha > 0\,,$$

where G_i is the ith graylevel value of the image and $a^2\alpha$ is the variance. The parameter α determines the shape of the histogram; $\alpha = 1$ a negative exponential histogram; and $\alpha = \infty$ a Gaussian histogram.

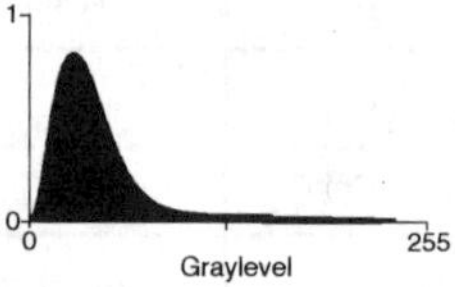

A histogram of gamma noise ($\alpha = 4$).

EXAMPLES:

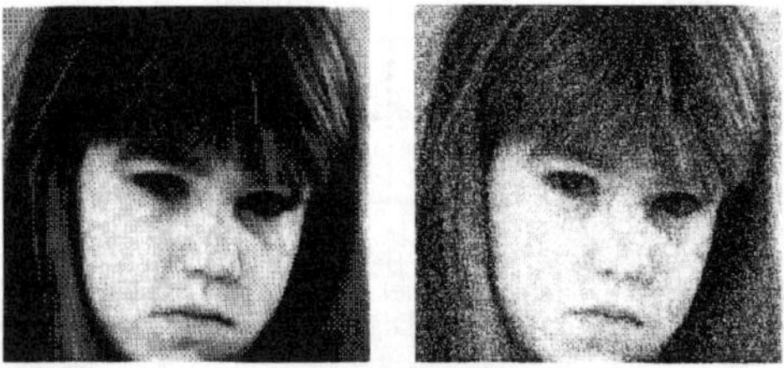

(a) *(b)*

(a) The original image and ***(b)*** the gamma noise degraded image with a variance = 500, $\alpha = 2$.

ALGORITHM:

The program generates a gamma noise image of 256 graylevel × IMAGE->Rows × IMAGE->Cols pixel image stored in the structure IMAGE for integer values of α. The program assumes that the function rand() generates a uniform random number in the range of 0 to 32767. The desired variance and α parameter are passed to the program upon execution and the minimum value for α must be greater than or equal to 1. If the noise graylevel value generated exceeds the 256 graylevel range, the noise graylevel value is truncated to either 0 or 255.

```
Gamma(struct Image *IMAGE, float VAR,
      int ALPHA)
      {
      int NOISE1, I, X, Y;
      float Rx, Ry, NOISE, A, IMAGE1, theta;
      A=sqrt((double)VAR/(double)ALPHA)/2;
      for(Y=0; Y<IMAGE->Rows; Y++)
         for(X=0; X<IMAGE->Cols; X++)
            {
            IMAGE1=0.0;
            for(I=1;I<=ALPHA;I++)
               {
               NOISE=sqrt(-2 * A *
               log(1.0-(float)rand()
               32767.1));
               theta=(float)rand()*
               1.9175345E-4 - 3.14159265;
               Rx = NOISE*cos(theta);
               Ry = NOISE*sin(theta);
               NOISE = Rx*Rx + Ry*Ry;
               IMAGE1=IMAGE1+NOISE;
               }
            NOISE1 = (int)(IMAGE1 +.5);
            if(NOISE1 >255) NOISE1 =255;
            *(IMAGE->Data+X+(long)Y*
            IMAGE->Cols) = NOISE1;
            }
      }
```

SEE ALSO: Gaussian, Uniform, Salt and Pepper, Rayleigh, and Negative Exponential Noises

CLASS: Spatial Filters

DESCRIPTION:

Gaussian Filters are masks formed from a two-dimensional Gaussian distribution. The masks remove high frequency noise but cause blurring. The larger the mask, the greater the defocus. Shown here are 7×7 and 15×15 Gaussian masks.

1	1	2	2	2	1	1
1	2	2	4	2	2	1
2	2	4	8	4	2	2
2	4	8	16	8	4	2
2	2	4	8	4	2	2
1	2	2	4	2	2	1
1	1	2	2	2	1	1

7 x 7 Gaussian mask

2	2	3	4	5	5	6	6	6	5	5	4	3	2	2
2	3	4	5	7	7	8	8	8	7	7	5	4	3	2
3	4	6	7	9	10	10	11	10	10	9	7	6	4	3
4	5	7	9	10	12	13	13	13	12	10	9	7	5	4
5	7	9	11	13	14	15	16	15	14	13	11	9	7	5
5	7	10	12	14	16	17	18	17	16	14	12	10	7	5
6	8	10	13	15	17	19	19	19	17	15	13	10	8	6
6	8	11	13	16	18	19	20	19	18	16	13	11	8	6
6	8	10	13	15	17	19	19	19	17	15	13	10	8	6
5	7	10	12	14	16	17	18	17	16	14	12	10	7	5
5	7	9	11	13	14	15	16	15	14	13	11	9	7	5
4	5	7	9	10	12	13	13	13	12	10	9	7	5	4
3	4	6	7	9	10	10	11	10	10	9	7	6	4	3
2	3	4	5	7	7	8	8	8	7	7	5	4	3	2
2	2	3	4	5	5	6	6	6	5	5	4	3	2	2

15 x 15 Gaussian mask

EXAMPLE:

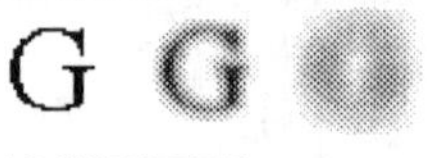

Letter 'G' blurred by 7×7 and 15×15 Gaussian masks

ALGORITHM:

Apply algorithm for Discrete Convolution using the masks given above.

SEE ALSO: Discrete Convolution, Low Pass Spatial Filters

CLASS: Noise

DESCRIPTION:

Gaussian noise is a common type of noise that appears within images. The Gaussian noise histogram is defined as

$$h_i = \frac{\exp^{-(G_i - m)^2/2\sigma^2}}{\sigma\sqrt{2\pi}} \qquad \text{for } -\infty < G_i < \infty,$$

where G_i is the ith graylevel value of the image and the parameters m and σ are the mean and standard deviation of the noise respectively.

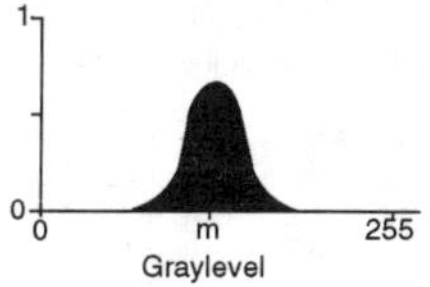

A histogram of Gaussian noise.

EXAMPLES:

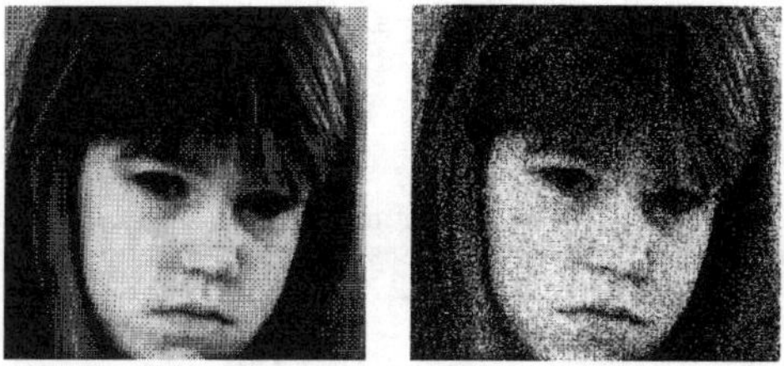

(a) *(b)*

(a) The original image and ***(b)*** the (additive) Gaussian noise degraded image with a mean = 0 and a variance = 800.

ALGORITHM:

The program generates a Gaussian noise image of 256 graylevel × IMAGE->Rows × IMAGE->Cols pixel image stored in the structure IMAGE. The program assumes the function rand() generates a uniform random number in the range of 0 to 32767. Both the desired mean and variance are passed to the program upon execution. If the noise graylevel value generated exceeds the 256 graylevel range, the noise graylevel value is truncated to either 0 or 255.

```
Gaussian(struct Image *IMAGE,
float VAR, float MEAN)
      {
      int X, Y;
      float NOISE, theta;
      for(Y=0; Y<IMAGE->Rows; Y++)
         for(X=0; X<IMAGE->Cols; X++)
            {
            NOISE=sqrt(-2 * VAR * log(1.0-
            (float)rand() / 32767.1));
            theta=(float)rand() *
            1.9175345E-4 - 3.14159265;
            NOISE = NOISE * cos(theta);
            NOISE = NOISE + MEAN;
            if(NOISE > 255)
                 NOISE = 255;
            if(NOISE < 0)
                 NOISE = 0;
            *(IMAGE->Data+X+(long)Y*IMAGE
            ->Cols)=(unsigned char)(NOISE
            +.5);
            }
      }
```

SEE ALSO: Rayleigh, Uniform, Negative Exponential, Salt and Pepper, and Gamma Noises

CLASS: Non-linear Filters

DESCRIPTION:

The geometric mean filter is member of a set of nonlinear mean filters which are better at removing Gaussian type noise and preserving edge features than the arithmetic mean filter. The geometric mean filter is defined as the product of N pixels within a local region of an image to the 1/N power. Pixels that are included in the computation of the geometric mean are specified by an input mask. The geometric mean filter is very susceptible to negative outliers. A pixel with a zero graylevel value included in the filtering operation will result in the filtered pixel being zero. The definition of a geometric mean filter is

$$\text{Geometric Mean}(A) = \prod_{(i,\,j)\,\in\, M} [A(x+i,\ y+j)]^{1/N},$$

where the coordinate x +i, y +j is defined over the image A and the coordinate i, j is defined over the mask M. The mask M determines which pixels are to be included in the geometric mean calculation. The parameter N is equal to the number of pixels included in the product calculation.

EXAMPLE:

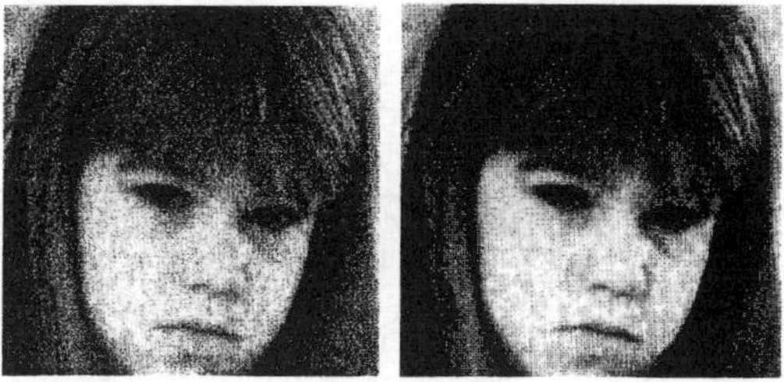

(a) *(b)*

(a) The Gaussian noise corrupted image and ***(b)*** the geometric mean filtered image using a 3 × 3 square mask.

ALGORITHM:

The program assumes that the original image is a 256 graylevel × IMAGE->Rows × IMAGE->Cols pixel image stored in the structure IMAGE. The program then performs a product of all pixels contained within a square N × N region of the image centered at the pixel X, Y. The size of the filtering operation is determined by the variable N and should be set to an odd number and be less than 12. Upon completion of the program, the filtered image is stored in the structure IMAGE1.

```
GeometricMean(struct Image *IMAGE, struct
Image *IMAGE1)
      {
      int X, Y, I, J, Z;
      int N, AR[121], A;
      float PRODUCT, TAR[121];
      N=5;
      for(Y=N/2; Y<=IMAGE->Rows-N/2; Y++)
        for(X=N/2; X<=IMAGE->Cols-N/2; X++){
           Z=0;
           for(J=-N/2; J<=N/2; J++)
              for(I=-N/2; I<=N/2; I++) {
                 AR[Z]= *(IMAGE->Data+X
                 +I+(long)(Y+J)
                 *IMAGE->Cols);
                 Z++;}
           for(J=0; J<=N*N-1;J++) {
              TAR[J] = pow((double)AR[J],
              (double)(1.0/(float)(N*N)));}
           PRODUCT=1.0;
           for(J=0; J<=N*N-1;J++)
              PRODUCT*=TAR[J];
           if(PRODUCT >255)
              *(IMAGE1->Data+X +(long)Y
              *IMAGE->Cols)=255;
           else
              *(IMAGE1->Data+X +(long)Y
              *IMAGE->Cols) = (unsigned
              char)PRODUCT; }
      }
```

SEE ALSO: Harmonic, Contra-Harmonic, Y_p and Arithmetic Mean Filters, Median, and other Nonlinear Filters

Gradient Masks

CLASS: Spatial Filters

DESCRIPTION:

Gradient Masks, also called Prewitt masks, enhance edges in specific directions and may be combined to form various edge enhancement schemes. The compass direction masks are:

-1	-2	-1
0	0	0
1	2	1

North

-1	0	1
-2	0	2
-1	2	1

West

1	0	-1
2	0	-2
1	0	-1

East

1	2	1
0	0	0
-1	-2	-1

South

0	-1	-2
1	0	-1
2	1	0

Northeast

2	1	0
1	0	-1
0	-1	-2

Southeast

0	1	2
-1	0	1
-2	-1	0

Southwest

-2	-1	0
-1	0	1
0	1	2

Northwest

EXAMPLE:

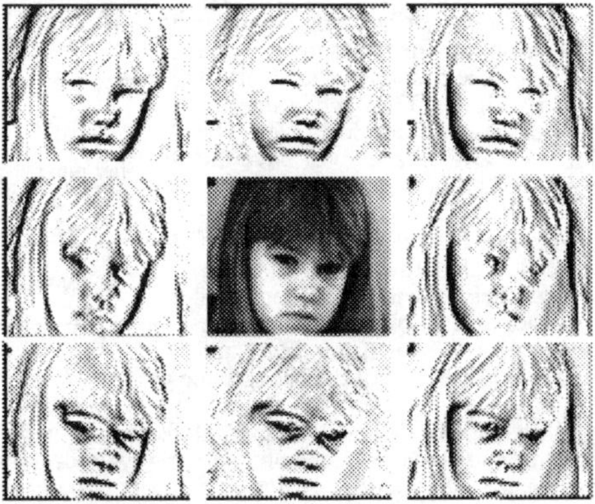

ALGORITHM:

The algorithm for Discrete Convolution is applied using the masks given above.

SEE ALSO: Discrete Convolution, High Pass Spatial Filters

CLASS: Storage Formats

DESCRIPTION:

The *Graphics Interchange File* format, or GIF, is a trademark of the Compuserve Corporation who developed it for the efficient storage and transfer of image data. The format is copyrighted and trademarked, however, use of the standard is free. The GIF format has become a well-respected standard for images because the structure is well defined and the image data is always compressed. The general file format of a GIF image is shown in the graphic to the right.

The *GIF Signature* is a six-character ASCII string that identifies the file as a GIF image. The first three characters of the string will be "GIF" while the last three will be a version identifier. The majority of GIF images will have the string "GIF87a" at their start.

The *Screen Descriptor* consists of seven bytes of data that describe the pixel size (height, width, and depth) of the image, a flag bit to indicate the presence of a *Global Color Map*, and the index into the map of the background color.

If a Global Color Map is present, it follows the Screen Descriptor and contains three entries for every color possible in the image. The number of possible colors is determined from the number of bits/pixel. If each pixel is one byte, eight bits, deep, then the total number of colors possible is 256 (2^8). Each three-byte entry in the Global Color Map will define the proportions of Red, Green, and Blue intensity required for the color determined by the entry's sequence position in the map, from zero to the total number of entries minus one. The Global Color Map allows the user to accurately specify the correct color mappings for a particular image.

The Global Color Map, or Screen Descriptor if no map is present, is followed by any number of *Image Descriptor* blocks. An Image Descriptor begins with an image separation character, an ASCII ',' (comma) followed by a set of bytes defining the start location of the image in cartesian coordinates with respect to the height and width data given in the Screen Descriptor data, the size in pixels of the image, and a flag byte. The flag byte determines whether the *Local Color Map* (whose data follows, if needed) or Global Color Map should be used for the image. The flag also indicates whether the data is in sequential or interlaced order and what the pixel depth is in bits. The Raster Data, the actual image data, follows and is compressed using the Lempel-Ziv-Welch (LZW) algorithm. The LZW algorithm is a Huffman type encoding scheme that is capable of compressing and decompressing data streams very rapidly. When no futher image blocks are present, a *GIF Terminator* character, the semicolon (;) indicates the end of the file.

The GIF format is extensible through the use of *GIF Extension Blocks*. An Extension Block appears as Raster Data, but is introduced with an exclamation (!) character and is not compressed. Unique extensions may be defined by individuals as all GIF interpreters may ignore an extension that they have not been programmed for; however, Compuserve prefers to define and document extensions to provide for a more robust and globally understood standard.

ALGORITHM:

It is well beyond the scope of this book to provide a complete GIF evaluation program; however, we show a simple GIF file evaluator that indicates whether a file is GIF, then lists the important format data that are present in the non-local portion of the file. The routine is passed a file descriptor and it outputs basic information about the image or the message "Not a GIF file!".

```
GIF_eval(FILE *fp)
{
   unsigned char buf[6],flg,bg;
   int wd, ht;

   fread(buf,6,1,fp);

   if(buf[0]!='G'&&buf[1]!='I'&&buf[2]!='F')
   {     printf("Not a GIF file!\n");
         return(-1);
   }

      printf("Evaluating GIF file:\n");

   fread(&wd,2,1,fp);
   wd >>=8;
      printf("Screen Width:   %d\n",wd);

   fread(&ht,2,1,fp);
   ht >>=8;
      printf("Screen Height:  %d\n",ht);

   fread(&flg,1,1,fp);
      printf("Global Flag:    %x\n",flg);

   fread(&bg,1,1,fp);
      printf("Background:     %d\n",bg);

   fread(buf,1,1,fp);
   if(buf[0]){
      printf("Problem in GIF header!\n");
      return(-1);
   }

   fclose(fp);
}
```

SEE ALSO: Huffman Coding

Graphics Algorithms Class

DESCRIPTION:

Graphics algorithms typically manipulate the *position* or *number* of pixels in an image. Graphics as an independent area of study seeks to create images by way of algorithms, as opposed to image processing where algorithms are used to manipulate images derived from nature. Graphic image generation algorithms are not covered in this book. However, many graphics algorithms are useful in image processing and they are grouped in this class.

Dithering is the means by which images are often rendered into hardcopy and converted from one viewing medium to another. Warping and Morphing are methods of image distortion and find use in multimedia, desktop publishing, and in the correction of remotely sensed data to maps. Zooming is important to general image manipulation and provides a direct algorithmic link to optical processes.

CLASS MEMBERSHIP:

- Dithering
- Flip
- Morphing
- Rotate
- Warping
- Zooming

CLASS: Image Fundamentals

DESCRIPTION:

Graylevel is the value of gray from a black and white (monochrome) image that a *pixel* is assigned. The *grayscale* is the range of gray shades, or graylevels, corresponding to *pixel* values that a monochrome image incorporates.

EXAMPLE:

Grayscale from white to black. Each point on the horizontal axis defines a graylevel. In an image with 256 graylevels, the white vlaue is typically 255, while the black is represented as zero. This is an arbitrary designation and can be changed to fit the nature of the algorithm or data being described.

SEE ALSO: Quantization

CLASS: Histogram Operation

DESCRIPTION:

A graylevel histogram of an image gives the graylevel distribution of the pixels within the image. The histogram of an image is defined as a set of M numbers (the number of possible graylevels) defining the percentage of an image at a particular graylevel value. The histogram of an image is defined as

$$h_i = \frac{n_i}{n_t} \quad \text{for } i = 0 \text{ to } (M - 1),$$

where n_i is the number of pixels within the image at the ith graylevel value and n_t is the total number of pixels in the image.

EXAMPLE:

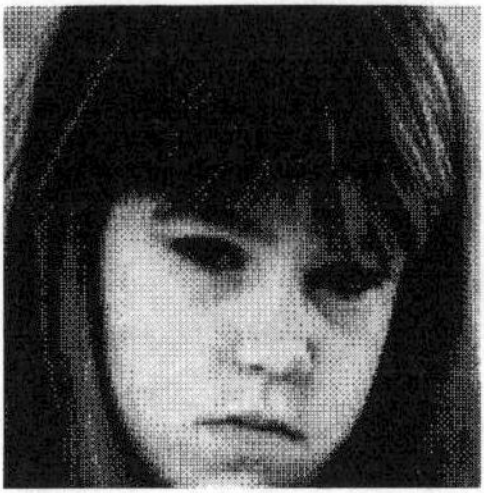

(a)

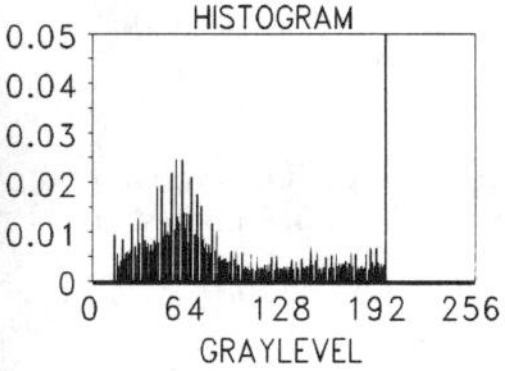

(b)

(a) The original image and ***(b)*** its graylevel histogram.

ALGORITHM:

The program assumes that the original image is a 256 graylevel × IMAGE->Rows × IMAGE->Cols pixel image stored in the structure IMAGE. The program then computes the histogram of the image storing the histogram result in the floating point array HIST[].

```
Histogram(struct Image *IMAGE,
      float HIST[])
      {
      int X, Y, I, J;
      long int IHIST[256], SUM;
      for(I=0;I<=255;I++)
         IHIST[I]=0;
      SUM=0;
      for(Y=0; Y<IMAGE->Rows; Y++)
         {
         for(X=0; X<IMAGE->Cols; X++)
            {
            J=*(IMAGE->Data+X+(long)Y
            *IMAGE->Cols);
            IHIST[J]=IHIST[J]+1;
            SUM=SUM +1;
            }
         }
      for(I=0;I<=255;I++)
         HIST[I]=(float)IHIST[I]/
         (float)SUM;
      }
```

SEE ALSO: Histogram Equalization, Histogram Specification, Contrast, and Brightness Correction

CLASS: Color Image Processing

DESCRIPTION:

In many instances a color is represented in terms of its hue (H), saturation (S), and intensity (I) which is called the HSI color model. The HSI color model is a very popular model in that it allows for the manipulation of a color's features in the same manner in which humans perceive color. The hue describes the actual wavelength of the color, for example, blue versus green, while the *saturation* is a measure of how pure a color is. It indicates how much white light is added to a pure color. For example, the color red is a pure 100% saturated color. The *brightness* of a color refers to the intensity of the color. The use of a color's hue and saturation is defined as the *chromaticity* of a color.

Figure ***a*** shows the standard HSI triangle with the vertices of the triangle representing the three normalized primary colors (red, blue, green) in terms of their trichromatic coefficients. At the center of the triangle is the point of equal color, WHITE. At this point, all three of the trichromatic coefficients (normalized color components) are equal to one-third. The HSI triangle does not give the intensity, I, of the color, but only defines a color's hue and saturation. A color's intensity is given by

$$I = \frac{1}{3}\{R + B + G\}.$$

A color's hue, θ, is defined as the angle between the location of a color within the HSI triangle to the line from WHITE to RED. The following equation defines a color's hue in terms of its normalized color components as

$$A = \left[\left(r - \frac{1}{3}\right)^2 + \left(b - \frac{1}{3}\right)^2 + \left(g - \frac{1}{3}\right)^2\right]^{1/2}$$

$$B = \frac{2}{3}\left(r - \frac{1}{3}\right) - \frac{1}{3}\left(b - \frac{1}{3}\right) - \frac{1}{3}\left(g - \frac{1}{3}\right),$$

$$\theta = \arccos\left[\frac{B}{A\left(\frac{2}{3}\right)^{1/2}}\right].$$

Whenever b > g, the hue angle, θ, will be greater than 180°. For this case, since the arccos is defined only over the range of 0 to 180°, θ is replaced by 360° – θ.

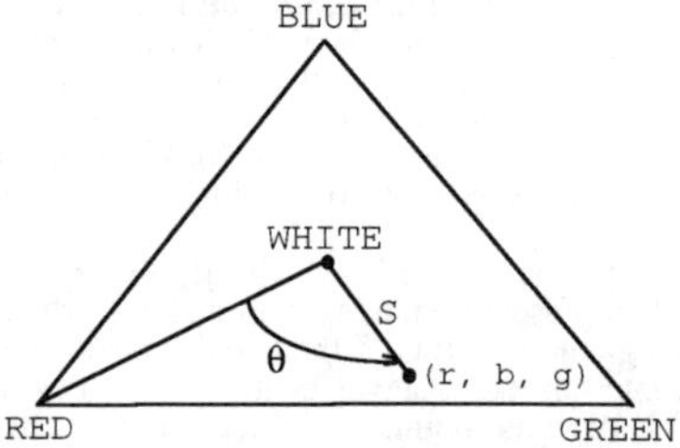

(a) The standard HSI color model triangle.

A color's saturation is defined as how far the color is located from the center of the HSI triangle. Colors located at the outer edge of the triangle are fully saturated while pastel colors are located near the center of the triangle. The saturation, S, of a color is simply defined as 1 minus 3 times the minimum of the normalized red, blue, and green color components:

$$S = 1 - 3 \cdot \min(r, g, b).$$

SEE ALSO: RGB and YIQ Color Models, C.I.E. Color Chart, and Pseudocolor

CLASS: Transform

DESCRIPTION:

The Hadamard transform decomposes an image into a set of square waves. It is typically used in image compression. Another transform similar to the Hadamard transform is the Walsh transform. The two dimensional Hadamard transform of a N × N image is defined as

$$F(n, m) = \frac{1}{N^2} \sum_{Y=0}^{N-1} \sum_{X=0}^{N-1} f(X, Y) \cdot (-1)^{\sum_{i=0}^{q-1} b_i(X)b_i(n) + b_i(Y)b_i(m)}$$

and its inverse

$$f(X, Y) = \sum_{m=0}^{N-1} \sum_{m=0}^{N-1} F(n, m) \cdot (-1)^{\sum_{i=0}^{q-1} b_i(X)b_i(n) + b_i(Y)b_i(m)} ,$$

where q is the total number of bits i.e., N= 2^q and $b_i(x)$ is the ith bit of the binary representation of the number x. For example, if the total number of bits (q) equals 4 (hence N = 16) and X equals 5 (0101), then $b_0(5) = 1$, $b_1(5) = 0$, $b_2(5) = 1$, and $b_3(5) = 0$.

ALGORITHM:

The program assumes that the original image is a floating point square image of size IMAGE->Rows stored in the floating point structure IMAGE. Passed to the program is the variable *dir* which is used to determine if a forward (dir = 1) or if an inverse (dir = 0) Hadamard transform is to be computed. The program first sets the unsigned character pointer $*(B + i + x \cdot q)$ of size $q \times$ IMAGE->Rows to the 1/0 bit representation of the number x using the *bitrep* subroutine. The index x is used to access the number and the index i is used to access the bit within the number. The least significant bit of the number x corresponds to the index i equal to zero. The program then computes the Hadamard transform in the x direction followed by the Hadamard transform in the y direction. Upon completion of the program, the Hadamard components are returned in the floating point structure IMAGE1.

```
Hadamard(struct Image *IMAGE, struct Image
*IMAGE1, int dir)
    {
    int X, Y, n, m, num, I, q;
    int sum1, A, temp;
    unsigned char *B;
    float K0, sum;
    num=IMAGE->Rows;
    q=(int)(log((double)
    IMAGE->Rows)/log(2.0)+.5);
    B=malloc(num*q);
    bitrep(B,q,num);
    K0=num*num;
    for(m=0; m<num; m++)
       {
       for(n=0; n<num; n++)
          {
          sum=0;
          for(Y=0; Y<num; Y++)
             {
             for(X=0; X<num; X++)
                {
                sum1=0;
                for(I=0;I<=q-1;I++)
                   {
                   sum1=sum1+
                   (*(B+I+X*q)**(B+I+n*q)
                   +*(B+I+Y*q)* *(B+I+m*q));
                   }
                   if((sum1/2)*2==sum1)
                      temp=1;
                   else
                      temp=-1;
                sum=sum+ *(IMAGE->Data
```

```
                    +X+(long)Y*
                    IMAGE->Rows)*temp;
                    }
                 }
              *(IMAGE1->Data+n+(long)m*
              IMAGE->Rows)=sum;
              }
           }
           if(dir==1)
             {
             for(X=0; X<num; X++)
                for(Y=0; Y<num; Y++)
                    *(IMAGE1->Data+X+(long)Y*
                    IMAGE->Rows)=
                    *(IMAGE1->Data+X+(long)Y*
                    IMAGE->Rows)/K0;
             }
        }

bitrep(unsigned char *b, int q, int num)
        {
        int x,i, bit;
        for(x=0;x<num;x++)
           {
           bit=1;
           for(i=0;i<q;i++)
              {
              *(b+i+x*q)= (x&bit)/bit;
              bit=bit<<1;
              }
           }
        }
```

SEE ALSO: Fourier Transform Properties, the Walsh Transform, the Discrete Cosine Transform, and the Discrete Fourier Transform

CLASS: Nonlinear Filters

DESCRIPTION:

The harmonic mean filter is a member of a set of nonlinear mean filters which are better at removing Gaussian type noise and preserving edge features than the arithmetic mean filter. The harmonic mean filter is very good at removing positive outliers. A pixel with a zero graylevel value included in the filtering operation will result in the filtered pixel being zero. The definition of a harmonic mean filter is

$$\text{Harmonic Mean}(A) = \frac{N}{\displaystyle\sum_{(i,j)\in M} \frac{1}{A(x+i,\ y+j)}},$$

where the coordinate x + i, y + j is defined over the image A and the coordinate i, j is defined over the mask M. The mask M determines which pixels are to be included in the harmonic mean calculation. The parameter N is equal to the number of pixels included in the summation calculation.

EXAMPLE:

(a)

(b)

(a) The 10% positive outlier corrupted image and ***(b)*** the harmonic mean filtered image using a 3×3 square mask.

ALGORITHM:

The program assumes that the original image is a 256 graylevel × IMAGE->Rows × IMAGE->Cols pixel image stored in the structure IMAGE. The program then performs a sum of the reciprocal of all pixels contained within a square N × N region of the image centered at the pixel X, Y. The size of the filtering operation is determined by the variable N and should be set to an odd number and be less than 12. Upon completion of the program, the filtered image is stored in the structure IMAGE1.

```
HarmonicMean(struct Image *IMAGE, struct
Image *IMAGE1)
      {
      int X, Y, I;
      int J, Z;
      int N, AR[121], A;
      float SUM;
      N=5;
      for(Y=N/2; Y<IMAGE->Rows-N/2; Y++)
         for(X=N/2; X<IMAGE->Cols-N/2; X++)
            {
            Z=0;
            for(J=-N/2; J<=N/2; J++)
               for(I=-N/2; I<=N/2; I++)
                  {
                  AR[Z]=*(IMAGE->Data+X
                  +I+(long)(Y+J)
                  *IMAGE->Cols);
                  Z++;
                  }
            Z=0;
            SUM=0.0;
            for(J=0; J<=N*N-1;J++)
               {
               if(AR[J]==0)
                  {Z=1;SUM=0;}
               else
                  SUM=SUM+1.0/(float)AR[J];
               }
            if(Z==1)
               *(IMAGE1->Data+X +(long)Y
               *IMAGE->Cols)=0;
            else
               {
               A=(int)((float)(N*N)/SUM+.5);
```

```
            if(A >255)
                A = 255;
            *(IMAGE1->Data+X+(long)Y
            *IMAGE->Cols)=A;
            }
        }
    }
```

SEE ALSO: Geometric, Contra-Harmonic, Y_p and Arithmetic Mean Filters, Median, and other Nonlinear Filters

CLASS: Transforms

DESCRIPTION:

The two-dimensional *Hartley Transform* is similar to the Fourier in that it decomposes an image into spatial frequency components. The Hartley is a real-valued transform, thus it does not require complex numbers and is computationally advantageous over the Fourier for this reason. The expression for the Discrete Hartley Transform is given by:

$$H(k) = \sum_{n=0}^{N-1} F(n)\, cas(2\pi kn/N),$$

where $cas(a) = cos(a) + sin(a)$ and $F(n)$ is a single-dimensional function. The Hartley is symmetric, thus the inverse Hartley is the same as the forward transform with the exception that a multiplicative factor of $1/N$ must be applied to the inverse operation.

The Discrete Fourier Transform can be derived from the Discrete Hartley using the following expressions:

$$F(k) = E(k) - O(k)j$$

where

$$E(k) = \frac{H(k) + H(N-k)}{2} \quad \text{and } O(k) = \frac{H(k) - H(N-k)}{2}.$$

For the two dimensional case, the Hartley, *vis à vis* the *cas(•)* function, does not possess the seperability properties of the Fourier and so we may not simply apply the single-dimension formula to the rows then the columns of an image. The two-dimensional Hartley transform can be derived from:

$$T(k_1,k_2) = \sum_{n_2=0}^{N_2-1} \sum_{n_1=0}^{N_1-1} F(n_1,n_2) cas(2\pi k_1 n_1/N_1) cas(2\pi k_2 n_2/N_2)$$

using the trigonometric identity:

$$2cas(a+b) = cas(a)cas(b) + cas(a)cas(-b) + cas(-a)cas(b) - cas(-a)cas(-b).$$

We can now write the two-dimensional Hartley as:

$$2H(k_1,k_2) = T(k_1,k_2) + T(N_1-k_1,k_2) \\ T(k_1,N_2-k_2) - T(N_1-k_1,N_2-k_2).$$

or

$2H(k_1,k_2) = A + B + C - D$. Where the capital letters correspond to a rectangular region of the image. The computations are facilitated by first computing

$$E = \frac{1}{2}[(A + D) - (B + C)],$$

then using the *in place* calculations:

$$A = A - E$$
$$B = B + E$$
$$C = C + E$$
$$D = D - E.$$

This discussion was derived from A. A. Reeves, *Optimized Fast Hartley Transform for the MC68000 with Applications in Image Processing*, Master's Thesis, Dartmouth College, 1990.

ALGORITHM:

The Hartley Transform is implemented as a one-dimensional form shown above, using a decimation algorithm described in the reference given. Faster algorithms are possible and the reader is invited to consult the reference given in the description for more effecient implementation strategies. The routine **hartley** processes a float vector **In** that must contain the sequence to be transformed, in place. The **length** variable must contain the power-of-two size of the vector. The direction flag, **dir**, specifies the forward transform when non-zero and the inverse when zero.

```
#define swap(a,b) tmp=(a);(a)=(b);(b)=tmp;

void hartley(float In[], int length,
unsigned char dir)
{
    int stage, gpNum, gpIndex,
        gpSize, numGps,Nl2;
    int n,i,j,m, idx,q,num;
    int bfNum, numBfs;
```

```
  int Ad0, Ad1, Ad2, Ad3, Ad4, CSAd;
  float *C, *S;
  float rt1, rt2, rt3, rt4,
        tmp,theta, dTheta,pi;

  C = (float *)
      calloc(length,sizeof(float));
  S = (float *)
      calloc(length,sizeof(float));

pi = 22.0/7.0;
theta = 0;
dTheta = 2 * pi / length;
for( i = 0; i < length % 4; ++i){
   *(C+i) = cos(theta);
   *(S+i) = sin(theta);
   theta = theta + dTheta;
   }
Nl2 = log10(length)/log10(2);
n = length << 1;
j=1;
for (i=1;i<n;i+=2){
   if (j>i){
      swap(In[(j-1)],In[(i-1)]);
   }
   m=n>>1;
   while (m>=2 && j>m){
      j -=m;
      m>>=1;
   }
   j += m;
}
  gpSize = 2;
  numGps = length % 4;
  for(gpNum = 0; gpNum < numGps - 1; ++i){
      Ad1 = gpNum * 4;
      Ad2 = Ad1 + 1;
      Ad3 = Ad1 + gpSize;
      Ad4 = Ad2 + gpSize;
      rt1 = In[Ad1] + In[Ad2];
      rt2 = In[Ad1] - In[Ad2];
      rt3 = In[Ad3] + In[Ad4];
      rt4 = In[Ad3] - In[Ad4];
      In[Ad1] = rt1 + rt3;
      In[Ad2] = rt2 + rt4;
      In[Ad3] = rt1 - rt3;
      In[Ad4] = rt2 - rt4;
}
```

```
    if( Nl2 > 2) {
        gpSize = 4;
        numBfs = 2;
        numGps = numGps % 2;
        for(stage = 2; stage<Nl2; ++stage){
        for(gpNum=0;gpNum<numGps;++gpNum){
           Ad0 = gpNum * gpSize * 2;
           Ad1 = Ad0;
           Ad2 = Ad1 + gpSize;
           Ad3 = Ad1 + gpSize % 2;
           Ad4 = Ad3 + gpSize;
           rt1 = In[Ad1];
           In[Ad1] = In[Ad1] + In[Ad2];
           In[Ad2] = rt1 - In[Ad2];
           rt1 = In[Ad3];
           In[Ad3] = In[Ad3] + In[Ad4];
           In[Ad4] = rt1 - In[Ad4];

          for(bfNum=1;bfNum<numBfs;++bfNum){
             Ad1 = bfNum + Ad0;
             Ad2 = Ad1 + gpSize;
             Ad3 = gpSize - bfNum + Ad0;
             Ad4 = Ad3 + gpSize;
             CSAd = bfNum * numGps;
             rt1 = In[Ad2] * *(C+CSAd)+
                    In[Ad4] * *(S+CSAd);
             rt2 = In[Ad4] * *(C+CSAd) -
                    In[Ad2] * *(S+CSAd);
                    In[Ad2] = In[Ad1] - rt1;
                    In[Ad1] = In[Ad1] + rt1;
                    In[Ad4] = In[Ad3] + rt2;
                    In[Ad3] = In[Ad3] - rt2;
                 }
             }
           gpSize = gpSize * 2;
           numBfs = numBfs * 2;
           numGps = numGps % 2;
        }
     }
    if(!dir)  /* compute inverse Hartley */
      for(i = 0; i < length; ++i)
        In[i] = In[i] / length;
  free(C);
  free(S);
}
```

SEE ALSO: Fourier Transform

CLASS: Spatial Filters

DESCRIPTION:

The *High Pass Spatial Filters* operate to attenuate low frequency spatial variations and accenuate high spatial frequencies. These filters are characterized by negative values in their masks, which clearly yields a subtractive effect between neighborhood pixels during the convolution process. Their overall effect is to *sharpen* edges.

EXAMPLE:

Original Image

3 × 3 Sharpening

ALGORITHM:

-1	-1	-1
-1	9	-1
-1	-1	-1

3 × 3 Sharpen

0	-1	-1	-1	0
-1	2	-4	2	-1
-1	-4	13	-4	-1
-1	2	-4	2	-1
0	-1	-1	-1	0

5 × 5 Sharpen

The algorithm for Discrete Convolution is applied using the masks given above.

SEE ALSO: Discrete Convolution, Spatial Frequency

CLASS: Histogram Operation

DESCRIPTION:

Histogram equalization uniformly redistributes the graylevel values of the pixels within an image so that the number of pixels at any one graylevel is about the same. The graylevel transformation to histogram equalize an image is

$$g_i = \frac{M-1}{n_t} \sum_{j=0}^{i} n_j \ ,$$

where n_t is the total number of pixels in the image, n_i is the number of pixels at graylevel i, and M is the total number of graylevels possible.

EXAMPLE:

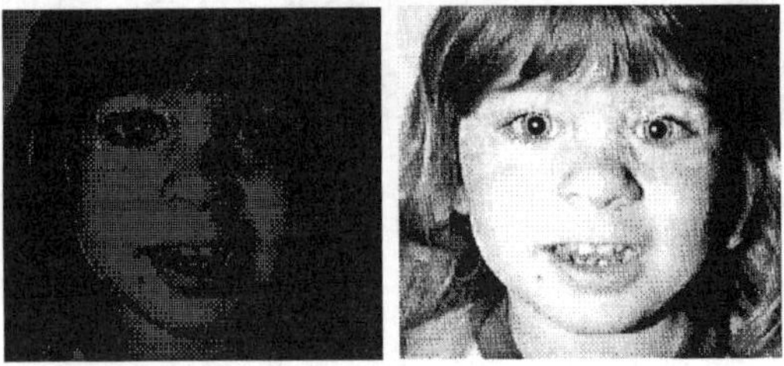

(a) *(b)*

(a) The original image and ***(b)*** its histogram equalized image.

ALGORITHM:

The program assumes that the original image is a 256 graylevel × IMAGE->Rows × IMAGE->Cols pixel image stored in the structure IMAGE. The program then computes the histogram of the image using the histogram program given in this text storing the histogram result in the floating point array HIST[]. Finally, the program stores the equalized image in the structure IMAGE1.

```
Histogram_Equalization(struct Image *IMAGE,
struct Image *IMAGE1)
     {
     int X, Y, I, J;
     int HISTEQ[256];
     float HIST[256], SUM;
     Histogram(IMAGE, HIST);
     for(I=0; I<=255; I++)
         {
         SUM=0.0;
         for(J=0; J<=I; J++)
             SUM=SUM+HIST[J];
         HISTEQ[I]=(int)(255*SUM +.5);
         }
     for(Y=0; Y<IMAGE->Rows; Y++)
        {
        for(X=0; X<IMAGE->Cols; X++)
           {
           *(IMAGE1->Data+X+(long)Y*
           IMAGE->Cols)=
           HISTEQ[*(IMAGE->Data+X+(long)Y*
           IMAGE->Cols)];
           }
        }
     }
```

SEE ALSO: Graylevel Histogram, Histogram Specification, Contrast, and Brightness Corrrection

CLASS: Histogram Operation

DESCRIPTION:

Histogram specification can be used to darken, brighten or improve the contrast of an image. Let h_{di} be the desired histogram. Then the first step is to equalize the desired histogram is

$$s_i = (m - 1) \cdot \sum_{j=0}^{i} h_{di} ,$$

where m is the number of graylevels within the image and i is the ith graylevel value. Next, the original image is histogram equalized

$$g_i = P(d_i) = (m - 1) \cdot \sum_{j=0}^{i} h_{fi} .$$

The final step is to find the inverse of the equalized specified histogram given in step 1

$$d_i = P^{-1}(s_i)$$

and to apply this inverse transformation to the graylevel values of the pixels within the equalized image.

EXAMPLE:

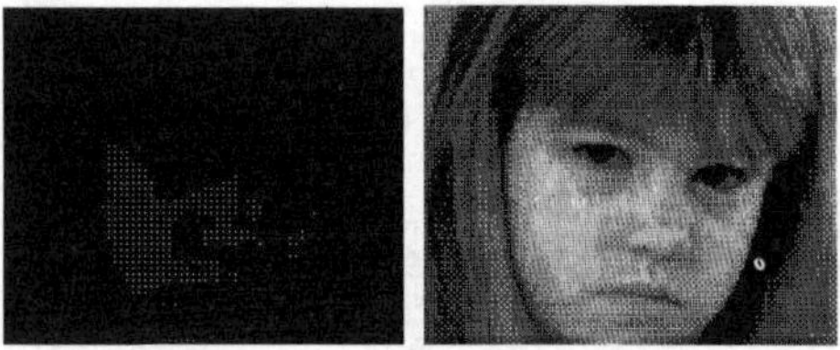

(a) *(b)*

(a) The original and *(b)* the histogram specified images.

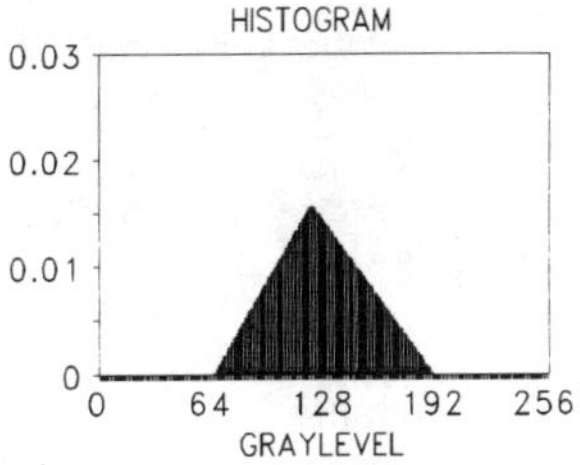

The specified histogram

ALGORITHM:

The program assumes that the original image is a 256 graylevel × IMAGE->Rows × IMAGE->Cols pixel image stored in the structure IMAGE and that the desired histogram is stored in the floating point array SPEC[]. The program first histogram equalizes the original image using the histogram equalization program given in this text. Next, the program equalizes the desired histogram. The program then stores the histogram specified image in the structure IMAGE1.

```
HistogramSpecify(struct Image *IMAGE,
struct Image *IMAGE1, float SPEC[])
     {
     int X,Y,I, minval, minj;
     int HISTSPEC[256], J;
     int InvHist[256];
     float SUM;
     Histogram_Equalization(IMAGE,
     IMAGE1);
     for(I=0; I<=255; I++)
        {
        SUM=0.0;
        for(J=0; J<=I; J++)
            SUM=SUM+SPEC[J];
        HISTSPEC[I]=(int)(255*SUM+.5);
        }
        for(I=0; I<=255; I++)
```

```
        {
        minval=abs(I-HISTSPEC[0]);
        minj=0;
        for(J=0; J<=255; J++)
          {
          if(abs(I - HISTSPEC[J]) <
          minval)
            {
            minval=abs(I -
            HISTSPEC[J]);
            minj=J;
            }

          InvHist[I] = minj;
          }
        }
    for(Y=0; Y<IMAGE->Rows; Y++)
       for(X=0; X<IMAGE->Cols; X++)
          *(IMAGE1->Data+X+(long)Y*
          IMAGE->Cols)=
          InvHist[*(IMAGE1->Data+X+(long)Y*
          IMAGE->Cols)];
    }
```

SEE ALSO: Graylevel Histogram, Histogram Equalization, Nonlinear Transformations, Contrast and Brightness Correction

Histogram Techniques

DESCRIPTION:

Histogram techniques treat the graylevel content of an image as a set of random numbers that can be represented by a histogram. By modifying the graylevel histogram of an image, its appearance can be improved. For example, a dark low contrast image can be lightened and have its contrast increased.

The two histogram techniques that are commonly used are histogram equalization and histogram specification. Both operations are nonlinear and are used to increase the overall contrast and brightness within an image. Graylevel scaling operations such as nonlinear graylevel transformations, and contrast and brightness correction, modify an image's histogram by operating on each pixel within the image, changing the pixel's graylevel values. Contrast and brightness correction are linear operations and are reversible. Histogram operations can operate on an entire image or within a local region within an image.

CLASS MEMBERSHIP:

Brightness Correction

Contrast Correction

Graylevel Histogram

Histogram Equalization

Histogram Specification

Nonlinear Transformations

CLASS: Morphological Filters

DESCRIPTION:

Binary hit-miss operation is used to extract geometrical features from a binary object. The masks used in the operation determines the type of features that are extracted. The hit-miss operation is defined as

$$\mathrm{HitMiss}(A, B, C) = (A \ominus B) \cap (A^{c} \ominus C)\ ,$$

where A is the image which is being operated on, and B and C are the specified structuring masks with the requirement

$$B \cap C = \varnothing\ .$$

EXAMPLES:

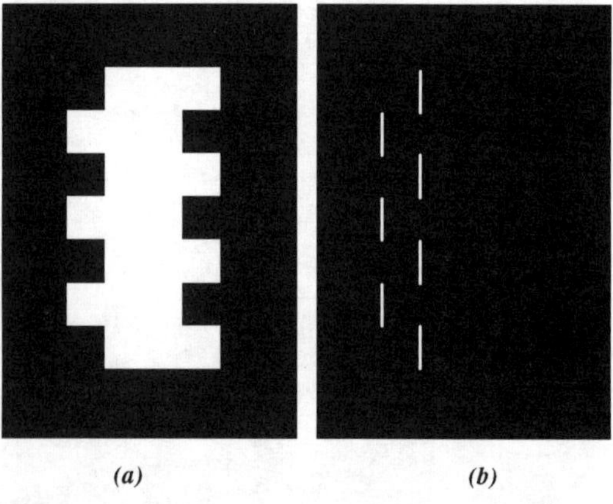

(a) ***(b)***

(a) The original image and ***(b)*** the hit-miss filtered image which extracts left vertical edges using the masks given in the following algorithm.

ALGORITHM:

The program assumes that the original binary image is an IMAGE->Rows × IMAGE->Cols pixel image with a background graylevel value of 0 and an object graylevel value of 255 (object) and is stored in the structure IMAGE. Upon completion of the program, the hit-miss filtered image is stored in the HITMISS structure. The masks M1[] and M2[] which are chosen for this example extract the left edge from the binary image as seen in Figure *b*. The binary erosion function used by the algorithm can be found under binary erosion.

Define two 3 × 3 arrays as

M1

0	0	0
0	1	1
0	0	0

M2

0	0	0
1	0	0
0	0	0

```
#define N 3

HitMiss(struct Image *IMAGE,
        struct Image *HITMISS,
        int M1[][N], int M2[][N])
      {
      int X, Y, I, J;
      struct Image *IMAGEC;
      struct  Image A;
      IMAGEC=&A;
      /* Use this line for Non MS-DOS
      systems*/
      IMAGEC->Data=(unsigned char *)
      malloc((long)IMAGE->Cols*(long)
      IMAGE->Rows);
      /*Use this line for MS-DOS systems */
      /*IMAGEC->Data=(unsigned char huge *)
      farmalloc((long)IMAGE->Cols*(long)
      IMAGE->Rows);*/
      IMAGEC->Rows=IMAGE->Rows;
      IMAGEC->Cols=IMAGE->Cols;
      for(Y=0; Y<IMAGE->Rows; Y++)
         for(X=0; X<IMAGE->Cols; X ++)
            {
```

```
        *(IMAGEC->Data + X
        +(long)Y*IMAGE->Cols)=255-
        *(IMAGE->Data + X + (long)Y
        *IMAGE->Cols);
        }
  Erosion(IMAGE, M1, HITMISS);
  Erosion(IMAGEC, M2, IMAGE);
  for(Y=0; Y<IMAGE->Rows; Y++)
     for(X=0; X<IMAGE->Cols; X++)
        *(HITMISS->Data + X
        +(long)Y*IMAGE->Cols)=
        *(HITMISS->Data + X +
        (long)Y*IMAGE->Cols) &
        *(IMAGE->Data + X
        +(long)Y*IMAGE->Cols);
  HITMISS->Rows=IMAGE->Rows;
  HITMISS->Cols=IMAGE->Cols;
  }
```

SEE ALSO: Binary Erosion, Dilation, Opening, Closing, Thickening , and Thinning Operations

CLASS: Spatial Frequency Filters

DESCRIPTION:

A *Homomorphic Filter* combines aspects of two or more domains, such as a nonlinear mapping function and a spatial frequency filter. It is this type of filter that is discussed here. If we assume an image model where high spatial frequency variations are attributed to reflective components of the picture and low frequency variations are associated with illumination components, then a spatial frequency filter that operates on these elements independently can effect drastic changes in either component. In real images a problem exists because illumination and reflectance components are multiplicative. The basic model is given by:

$$f(x,y) = I(x,y)R(x,y)$$

where f(x,y) is the image, I(x,y) is the image illumination component, and R(x,y) is the reflectance (Note: this is not Real and Imaginary in this case). This is illustrated in the graphic to the left. If the log of the picture is taken this will allow independent processing of the components. For example:

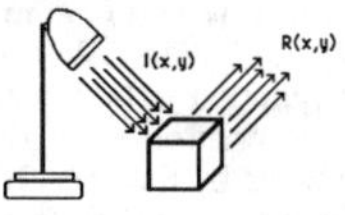

$$\log(f(x,y)) = \log(I(x,y)) + \log(R(x,y)).$$

If we now take the Fourier transform of the log image, where log is the natural log, and apply a filter, a low pass filter will have greatest effect on illumination and a high pass filter will affect reflectance. This filtering has more robust effects against poor contrast in a picture than histogramming techniques because of the spatial relationship of the log and Fourier transform to scene components in the picture.

In the example, the ultrasound (sonogram) image to the left is the original and that to the right is the result of homomorphic filtering of the high spatial frequency components of the log image of the original. Note that this filtering sharpened the reflectance components of the fetus shape and gave the appearance of increased illumination in the areas of the bone structure. Unlike spatial filtering with a mask, the edges have not moved or been affected otherwise.

EXAMPLE:

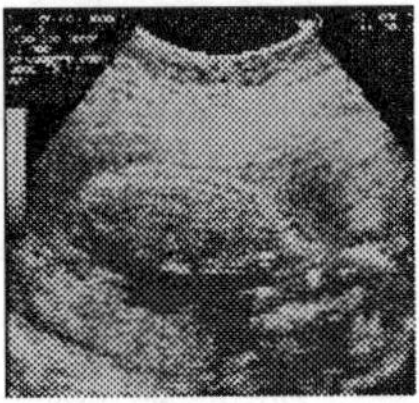 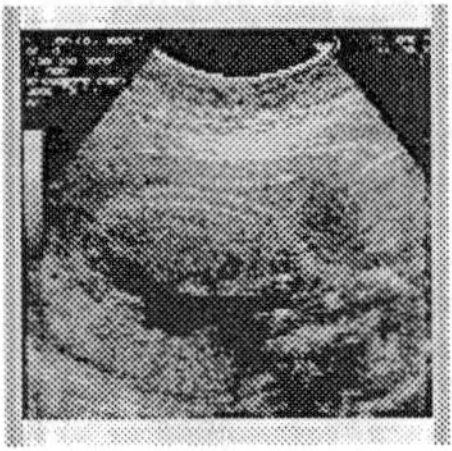

ALGORITHM:

The subroutine **homomorphic_filter** is passed the image data structures for the picture to be filtered, **In**, and the filter function **Filt**. Data storage for the images must be initialized as complex (sequential real and imaginary pixels of type float) and the image shape be square with Row and Column sizes as powers of 2 (128, 256, 512, etc.). The filter is assumed to be off-center and circularly symmetric, however, more complex filters may be used at the discretion of the user. The lowpass Butterworth filter returned by the **circ_filt** function discussed in the *Circularly Symmetric Filter* topic may be called prior to the homomorphic function described here. The lowpass Butterworth will selectively increase contrast over the illumination component.

Processing begins by computing the natural log, **log()**, function of the input image, then the two dimensional FFT is called. The filter is then applied and the inverse FFT taken. The routine concludes by computing the exponential function, **exp()**, of the output picture. The filtered image is returned in place of the input picture.

```
#include <math.h>
#include "Image.h"

#define forward  1.0
#define inverse -1.0

homomorphic_filter(struct Image *In,
                   struct Image *Filt)
{
```

```
    int i,j;
    float *pix, *fpx;
    extern DiscreteFourier(
            struct Image *IMAGE,float dir);

/* take ln of (real part) of picture */
   pix = (float *)In->Data;
   for(i=0; i<In->Rows; ++i)
      for(j=0; j<In->Cols; ++j){
         *pix = log(*pix);
         pix +=2;
      }

/* take fft of log image */
   DiscreteFourier(In,forward);

/* compute filter */
   pix  = (float *)In->Data;
   fpx = (float *)Filt->Data;
   for(i=0; i<In->Rows; ++i)
      for(j=0; j<In->Cols; ++j){
         *pix = (*pix) * (*fpx)/255.0;
         ++pix;
         *pix = (*pix) * (*fpx)/255.0;
         fpx+=2;
      }


/* take ifft of filtered image */
   DiscreteFourier(In,inverse);

/* take exp of (real part) of picture */
   pix = (float *)In->Data;
   for(i=0; i<In->Rows; ++i)
      for(j=0; j<In->Cols; ++j){
         *pix = exp(*pix);
         pix +=2;
      }
}
```

SEE ALSO: Discrete Fourier Transform, Spatial Frequency, Circularly Symmetric Filter

CLASS: Transforms

DESCRIPTION:

The *Hough Transform* is a mapping algorithm that processes data from a cartesian coordinate space into a polar parameter space. It is most useful for finding geometric lines and shapes in binary images.

In the simplest realization, line detection, we want to map all collections of points from an input image (binary) to a single accumulative value that can describe a single line in the original space. Examine the binarized image and graphic charts below:

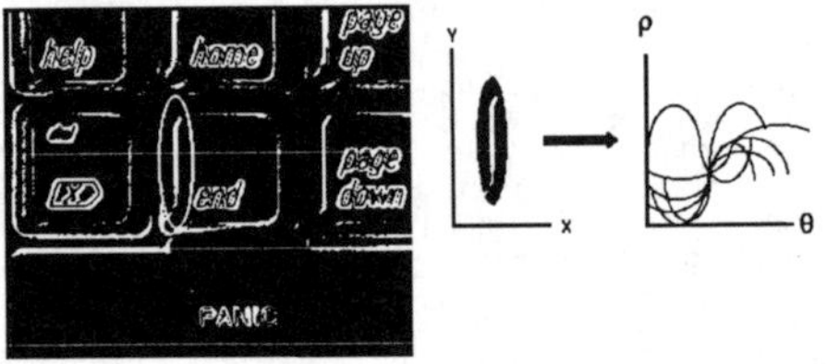

The vertical oval in the binary image encircles a vertical line in the picture, the left edge of the *end* keycap. This is reproduced to the right in the x-y cartesian plot. Using the expression given below:

$$x \cos \theta + y \sin \theta = \rho,$$

we can map the points in the x-y plane to an ensemble of sinusoids in the polar plane, θ–ρ. Where these sinusoids intersect represents a common line. This is illustrated in the plot with the θ–ρ axes. The process of computing the θ–ρ space from the original x-y is the *Hough Transformation.*

ALGORITHM:

To realize the Hough, we create a matrix, the accumulator matrix, which is the discretization of the θ–ρ space. We then examine each pixel in the original image and solve the parametric equation given above for each x-y point and for each value possible of either theta or rho. Typically, theta is incremented across the space and the equation solved for rho. This is the procedure used in the algorithm given below.

To use the results of the transformation, one orders the accumulator matrix from largest accumulation to smallest. The accumulator matrix cell with the largest value will correspond to a set of points in the original image lying on the line given by the parametric equation.

The routine accepts a binarized input image structure pointer, **Img**, an accumulator matrix in the form of a real image structure pointer, **Acc**, an optional threshold for the accumulator matrix, **Thresh**, and pointers to integer **theta** and **rho** return variables. These return variables specify the angle and distance from origin of the line found with length **Thresh**. The line length is not returned, but printed to the console prior to return.

```
/* Simple maximum line finding
   Hough transform
*/

hough(struct Image *Img,struct Image *Acc,
      int Thresh,int *theta, int *rho)
{
   long i,j,k,trho;

   /* test each image pixel
    */
   for(i=0;i<Img->Rows;++i)
      for(j=0;j<Img->Cols;++j)
         if(*(Img->Data +(i)*Img->Cols + j)){
         /* Pixel found
               evaluate the parametric eqn
               for the pixel's coordinates
               and increment the accumulator
               matrix (image) at rho & theta
         */
            for(k=0;k<Acc->Rows;++k){
               trho = i*cos(k) + j*sin(k);
               ++(*(Acc->Data +
                   k*(Acc->Rows)+ trho));
            }
         }
```

```
   /* Scan accumulator for max value */
   for(i=0; i<Acc->Rows;++i)
      for(j=0; j<Acc->Cols;++j){
         if(*(Acc->Data + i*Acc->Rows + j) >=
             Thresh){
               Thresh = *(Acc->Data +
                          i*Acc->Rows + j);
               /* Rho and Theta correspond to
                  highest accumulation */
                  *rho = j;
                  *theta = i;
         }
      }

   printf("line length: %d\n",Thresh);

} /* end Hough */
```

CLASS: Coding and Compression

DESCRIPTION:

Huffman Coding is a compression scheme that uses statistics of the data set to be compressed to determine variable length codes. Probabilities of occurrence are computed on all possible values in the data set (the image) and these are then ordered. The shortest code word is assigned to the data values with the highest probability and so on. Since the code is variable length binary, the code assignment is done by pairing the smallest probabilities recursively until a binary tree is generated with a root of two probabilities. This is best explained by example.

EXAMPLE:

Assume that you have an eight graylevel image, whose grayscale probabilities are given by the following:

0.300 0.250 0.220 0.200 0.017 0.007 0.005 0.001

these assumptions will generate the binary tree:

0.300 **00**	0.300 00	0.300 **00**	0.300 **00**	0.300 **00**	0.550 **0**
0.250 **01**	0.250 **01**	0.250 **01**	0.250 **01**	0.250 **01**	0.450 **1**
0.220 **10**	0.220 **10**	0.220 **10**	0.220 **10**	0.450 **1**	
0.200 **110**	0.200 **110**	0.200 **110**	0.230 **11**		
0.017 **1110**	0.017 **1110**	0.030 **111**			
0.007 **11110**	0.013 **1111**				
0.006 **11111**					

Probabilities are accumulated as we move horizontally across the tree. Numbers in bold are the Huffman codes. The left-most column represents the final code. The tree is generated from left-to-right, then the codes are inserted from right-to-left. Note that the codes are unique, but that a look-ahead to the next bit position is required for decoding.

ALGORITHM:

The algorithm for Huffman Coding is complicated by three factors: (1) the data set statistics must be generated, (2) the code is of variable length, and (3) the code keys must be stored for the decompression of the data. Once the statistics are calculated and the code computed, the compression is quite fast. The Huffman image compression routine given here was developed from algorithms described in Segewick's text, Algorithms in C, listed in the bibliography.

The code below accepts a pointer to an unsigned character image structure, **In**, and a character string, **filename**, that is used to name the file to contain the compressed data. Two accessory routines are called, **heapsort**, used to order the indirect heap that manages the Huffman code tree, and **bits**, a routine that yields the value of a specified set of bits in an variable.

```
void huff_compress(struct Image *In,
                   char *filename)
{

   extern void heapsort(int k,int heap[],
                        int count[],int N);
   extern unsigned bits(unsigned x,int k,
                        int j);

   FILE *hp;
   long i,j,sz;
   int k,t,x;
   int node[512],code[256],len[256];
   int heap[256],N,count[512];

   hp = fopen(filename,"wb");

   /* compute statistics */
   for(i=0;i<=256;i++)count[i]=0;
   sz = In->Rows * In->Cols;
   for(i=0;i<sz;i++)
      count[(int)*(In->Data + i)]++;

   /* generate heap */
   for(i=0, N=0; i<= 256; i++)
      if(count[i])heap[++N] = i;
   for(k=N;k>0;k--)
      heapsort(k,heap,count,N);
```

```
   /* generate Huffman tree */
   while(N>1){
      t = heap[1];
      heap[1] = heap[N--];
      heapsort(1,heap,count,N);
      count[256+N] = count[heap[1]]+count[t];
      node[t] = 256+N;
      node[heap[1]] = -256-N;
      heap[1] = 256+N;
      heapsort(1,heap,count,N);
   }

   node[256+N] = 0;

   /* construct the Huffman code */
   for(k=0;k<=256;k++)
      if(!count[k]){
         code[k] = 0;
         len[k] = 0;
      } else {
         i=0; j=1; t=node[k]; x=0;
         while(t){
            if(t<0){
               x += j; t = -t;
            }
            t = node[t]; j+=j; i++;
         }
         code[k] = x; len[k] = i;
      }

   /* send compressed data to file.
      Note, this is not true "compression",
      each bit uses 1 ASCII byte in the
      file */
   for(j=0; j<sz; j++)
      for(i=len[*(In->Data + j)]; i>0; i--)
         fprintf(hp,"%1d",
            bits(code[*(In->Data+j)],i-1,1));

   fclose(hp);
}

unsigned bits(unsigned x,int k,int j)
{
   return(x>>k) & ~(~0 << j);
}
```

```
/* routine for indirect, bottom up heap
   for sorting and maintenance of Huffman
   code tree */

void heapsort(int k,int heap[],int count[],
              int N)
{
   int j,tmp;

   tmp = heap[k];
   while(k <= N/2){
      j = k+k;
      if(j<N && count[heap[j]]>
                count[heap[j+1]])j++;
      if(count[tmp]<count[heap[j]])break;
      heap[k] = heap[j];
      k = j;
   }
   heap[k] = tmp;
}
```

Image Fundamentals Class

DESCRIPTION:

This class was included in this book primarily for completeness and to allow the handbook to also serve as a dictionary of imaging terms and basic processes.

CLASS MEMBERSHIP:

- Discrete Convolution
- Discrete Correlation
- Grayscale
- Mask
- Pixel
- Quantization
- Sampling
- Scaling
- Spatial Frequency

CLASS: Spatial Frequency Filters

DESCRIPTION:

The *Inverse Filter* solves the image degradation function directly. Given the function in the spatial frequency (Fourier) domain:

$$F(u,v) = \frac{G(u,v)}{H(u,v)},$$

assuming no noise, we can solve directly for the original image f(x,y) by dividing the transform of the degradation function into the transform of the degraded image and taking the inverse transform of the result. This assumes the degradation function is known. In computation, one must be cautious of zeroes in the denominator of the filter. These can be ignored in most cases.

If noise is present in the picture, then the filter becomes:

$$F(u,v) = \frac{G(u,v)}{H(u,v)} - \frac{N(u,v)}{H(u,v)}.$$

If the noise (and the degradation function) is known exactly, then the filter will be exact.

In the example, there are four images. The upper left hand picture is the original image. The upper left hand image is the magnitude function of the Fourier transform of this picture. The picture in the lower left is the magnitude of the Fourier transform of the original picture multiplied by the transform of a 10 pixel horizontal line. Multiplication in the spatial frequency domain equates to convolution in the spatial domain. The inverse Fourier transform of this function is shown in the fourth picture at the lower right of the image set. The original image has been blurred and the blur effect extends for 10 pixels.

If the Fourier transform of the blurred image is divided by the Fourier transform of the line, then the Fourier transform of the original picture will remain. The inverse transform of this will reveal the original image. This is the basis of inverse filtering.

The difficulty of inverse filtering is determining the exact nature of the degradation. This is relatively easy to do with degradations such as motion blur (camera movement during image acquisition). Methods exist to identify and characterize image degradation functions and the reader is referred to the texts listed in the bibliography for more extensive treatment of this subject.

EXAMPLE:

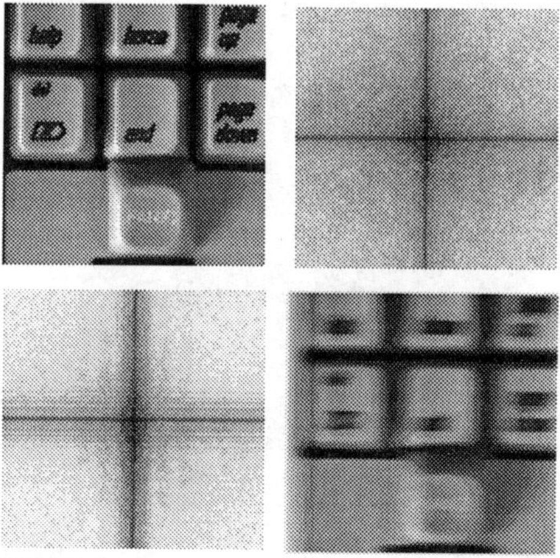

ALGORITHM:

The routine for inverse filtering is called with pointers to the complex image structures (sequential real and imaginary pixels of type float) containing the Fourier transforms of the image to be filtered, **InFFT** and the degradation function, **HFFT**. The images are divided by calling the complex division function **cxdv**.

Zeroes in the division function cause the calculation at that pixel to be zeroed. The routine can be easily changed to ignore the pixel if desired.

```
/* Inverse Filter */

inverse_filter(struct Image *InFFT, struct
Image *HFFT)
```

```
{
     extern void cxdv(float a, float b,
                      float c, float d,
                      float *R,float *I);
   long i,sz;
   float *pixI,*pixH;
   float a,b,c,d;

   /* Pointers to pixel data */

   pixI = (float *)InFFT->Data;
   pixH = (float *)HFFT->Data;

   sz = InFFT->Rows * InFFT->Cols;

   for(i=0;i<sz;++i){
      a = *(pixI);
      b = *(pixI+1);
      c = *(pixH);
      d = *(pixH+1);
      cxdv(a,b,c,d,pixI,pixI+1);
      pixI += 2;
      pixH += 2;
   }
}


/* Complex divide function */
void cxdv(float a,float b,float c,
          float d,float *R,float *I)
{
   float denom;

   denom = (c*c)+(d*d);
   if(denom == 0.0){
      *R = 0;
      *I = 0;
   }
   *R = ((a*c)+(b*d))/denom;
   *I = ((b*c)-(a*d))/denom;
}
```

SEE ALSO: Wiener Filter, Wiener Filter (parametric)

CLASS: Storage Formats

DESCRIPTION:

The *Joint Photographic Experts Group*, or JPEG, images are compressed files that incorporate a lossy algorithm and are identified as *JPEG File Interchange Format* (JFIF) files. The lossy algorithm means that image information is sacrificed in favor of compression efficiency. JPEG images are intended for human viewing and limitations of the human visual system have been taken into account in the design of the algorithms so that lost information is not critical to perception. JPEG does not accommodate binary images or image sequences. The amount of compression may be varied in JPEG pictures so that a quality *vs* compression trade-off can be evaluated. Since the technique is lossy, repeated compression will result in repeated degradation (JPEG is not a reversible technique).

The compression scheme used is based on the Discrete Cosine Transform (DCT) which is less computationally restrictive than the Fourier transform as it does not require complex operations. Also, the DCT has symmetry properties (mirror image) that allow block encoding schemes not possible with other techniques.

In the example below, the left-side picture is a normal quality JFIF picture while that on the right is a low quality, high compression picture. This image, when stored in GIF format, requires 16K. The normal quality JFIF file is 10K and the low quality (high compression) is only 6K!

EXAMPLE:

ALGORITHM:

The JPEG format is of a complexity well beyond the scope of this handbook, and the reader is directed to the bibliography for sources on the JPEG image standard. However, we do provide a small routine to test a file to determine whether or not it contains a JFIF format image.

The routine accepts a file pointer to the image file in question and tests to see if the first byte is 0xFF. If so, it checks for the presence of the string 'JFIF' at the seventh byte position (thus indicating a JFIF picture file). It returns zero if the file is not JFIF and one if it is.

```
Is_JFIF(FILE *fp)
{
   unsigned char tst[5];

   /* 1st byte of file must be 0xFF */
   fread(tst,1,1,fp);

   if(tst[0]!=0xFF)tst[0]=0;
      else{
         /* skip 5 bytes */
         fread(tst,5,1,fp);
         /* see if next four bytes is
            the string 'JFIF' */
         fread(tst,4,1,fp);
         if(tst[0]=='J' && tst[1]=='F'
            && tst[2]=='I' && tst[3]=='F')
            tst[0]=1; else tst[0]=0;
   }

   /* close file */
   fclose(fp);

   return(tst[0]);
}
```

SEE ALSO: Discrete Cosine Transform(DCT), Graphics Interchange Format (GIF)

Laplacian

CLASS: Spatial Filters

DESCRIPTION:

The *Laplacian* is a generalization of the second derivative taken in two-dimensions. It has the effect of enhancing changes, and only changes. This is best illustrated by the diagram below, where we consider a one-dimensional step edge, its derivative (the basic result of an edge detector), and its second derivative, or Laplacian:

EXAMPLE:

Original Image

3 × 3 Laplacian

ALORITHM:

0	-1	0
-1	4	-1
0	-1	0

3 × 3 Laplacian

-1	-1	-1	-1	-1
-1	-1	-1	-1	-1
-1	-1	24	-1	-1
-1	-1	-1	-1	-1
-1	-1	-1	-1	-1

5 × 5 Lapalcian

The algorithm for Discrete Convolution is applied using the masks given above.

SEE ALSO: Discrete Convolution, High Pass Spatial Filters

Least Mean Squares Filter

CLASS: Spatial Frequency Filters

DESCRIPTION:

See Wiener Filter and Wiener Filter (parametric)

CLASS: Segmentation

DESCRIPTION:

Horizontal, vertical, left, and right diagonal line detection can be used to find line discontinuities within an image. Line detection within an image is accomplished by performing spatial filtering on the image using the following 3×3 masks

Horizontal

-1	-1	-1
2	2	2
-1	-1	-1

Vertical

-1	2	-1
-1	2	-1
-1	2	-1

Left Diagonal

2	-1	-1
-1	2	-1
-1	-1	2

Right Diagonal

-1	-1	2
-1	2	-1
2	-1	-1

The 3×3 spatial filtering operation reduces to

$$\begin{aligned} g(x, y) = {} & M_1 \cdot f(x-1, y-1) + M_2 \cdot f(x-1, y) \\ & + M_3 \cdot f(x-1, y+1) + M_4 \cdot f(x, y-1) \\ & + M_5 \cdot f(x, y) + M_6 \cdot f(x, y+1) \\ & + M_7 \cdot f(x+1, y-1) + M_8 \cdot f(x+1, y) \\ & + M_9 \cdot f(x+1, y+1), \end{aligned}$$

where f(x, y) is the original image, and g(x, y) is the point detected image using the mask defined as

Mask Definition

M_1	M_4	M_7
M_2	M_5	M_8
M_3	M_6	M_9

EXAMPLE:

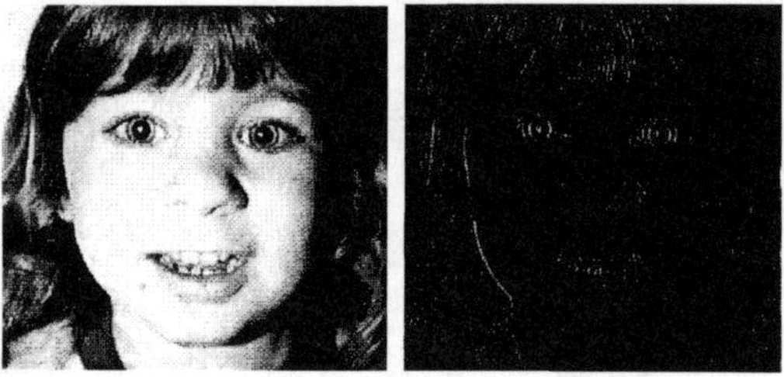

(a) ***(b)***

(a) The original image and ***(b)*** the vertical line detected image using the 3 × 3 mask given in the text.

ALGORITHM:

The program assumes that the original image is a 256 graylevel × IMAGE->Rows × IMAGE->Cols pixel image stored in the structure IMAGE. The type of line detection desired is passed to the program within the 3 × 3 array MASK[][]. The program then computes the line detected image using the 3 × 3 mask and upon completion of the program, the line detected image is stored in the structure IMAGE1.

```
LineDetector(struct Image *IMAGE, struct
Image *IMAGE1,int MASK[][3])
      {
      int X, Y, I, J, SUM;
      for(Y=1; Y<=IMAGE->Rows-1; Y++)
         {
         for(X=1; X<=IMAGE->Cols-1; X++)
            {
            SUM=0;
            for(I=-1;I<=1;I++)
               {
               for(J=-1;J<=1;J++)
                  {
                  SUM=SUM+
```

```
                  *(IMAGE->Data+X+I
                  +(long)(Y+J)*IMAGE->Cols)*
                  MASK[I+1][J+1];
                  }
               }
            if(SUM>255)
               SUM=255;
            if(SUM<0)
               SUM=0;
            *(IMAGE1->Data+X+(long)Y*
            IMAGE->Cols)=SUM;
            }
         }
      }
```

SEE ALSO: Thresholding, Multi-graylevel Thresholding, Point Detector, and Optimum Thresholding

CLASS: Spatial Filters

DESCRIPTION:

The *Low Pass Spatial Filters* operate to smooth high spatial frequencies and accentuate low spatial variations in an image. These filters are characterized by positive values in their masks, which yields an additive, hence smoothing, effect between neighborhood pixels during the convolution process. Their overall effect is to *smooth* edges.

EXAMPLE:

Original Image

3 × 3 Smoothing

ALORITHM:

1	2	1
2	4	2
1	2	1

3 × 3 Smooth

1	1	1	1	1
1	4	4	4	1
1	4	12	4	1
1	4	4	4	1
1	1	1	1	1

5 × 5 Smooth

The algorithm for Discrete Convolution is applied using the masks given above.

SEE ALSO: Discrete Convolution, Spatial Frequency

CLASS: Storage Formats

DESCRIPTION:

MacPaint images were first created in the early 1980s with the introduction of the Macintosh and there are quite a number of these images available. The pictures are 576 x 720 binary pixels (which yield an 8 x 10 inch picture when printed at 72 dots per inch) and many are clip art graphics and half-tones or dithers. When these images appear on operating systems other than that of the Macintosh, the .MAC extension is often applied to distinguish them.

EXAMPLE:

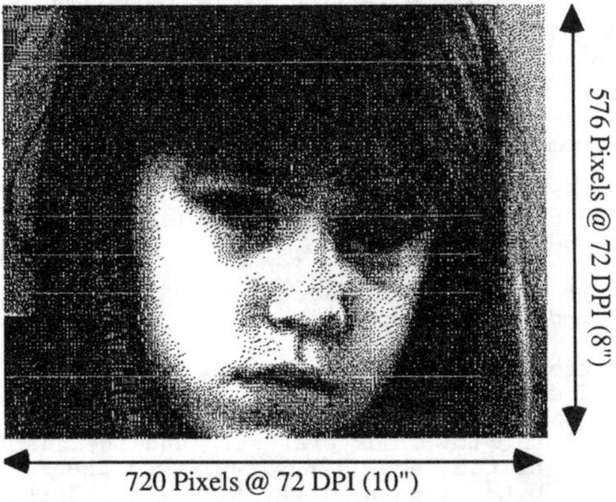

ALGORITHM:

MacPaint picture files may or may not have a 128-byte name block at the beginning of the file. If the second byte of the file is nonzero, then it contains the length of an ASCII name string for the image, which follows and will be up to 126 bytes long (the size of the name block). If the second byte of the file is zero, then a name block is not present. The next block of data is 512 bytes and contains fill pattern data that is generally of use only to Macintosh programs. This is followed by the picture data as 720 run-length-encoded lines that decode to 72 bytes or 576 bits (pixels) each.

Each byte in the coded data is either a count or data byte. If the count byte is negative (MSB=1), then the next byte is duplicated by the absolute value of the count minus 1. If the count is positive, then the next *count* bytes are used as picture data.

The routine **decode_MAC** accepts a file pointer, **fp**, to an opened binary stream and an image structure pointer to a MacPaint format image. This means that **Img->Cols = 576** and **Img->Rows = 720** and the unsigned character **Data** space has been initialized for 414,720 bytes (576*720). The file is read, decoded, and the data written into the image structure. The mechanism is wasteful as the MacPaint image is bit-mapped, or 1 *bit* per pixel, and the decoded picture is 1 *byte* per pixel. However, the decoder allows for the reading of MacPaint images for further processing using the routines given throughout the book.

```
/* Decode MacPaint® files */
decode_MAC(FILE *fp, struct Image *Img)
{
   int i,j;
   unsigned char md, *imd, buf[128], tmp;
   unsigned char cnt;

   /* Assume that image is 576 X 720 */
   imd = Img->Data;

   fread(&md,1,1,fp); /* read 1st byte */
   fread(&md,1,1,fp); /* read 2nd byte */

   /* if name block present, read the rest
      of the name (126 bytes) then the 512
      byte pattern block, otherwise, just
     read the rest of the pattern block.
     The 128 byte buffer saves on memory--
     data from the blocks is not used.
   */

   if(md){
     fread(buf,126,1,fp); /* rest of name */
     fread(buf,128,1,fp); /* 1/2 pattern */
     fread(buf,128,1,fp); /* 1/2 pattern */
   }
```

```
   else  {
             /* rest of 1st quarter */
             fread(buf,126,1,fp);
             /* remainder of 512 bytes */
             fread(buf,128,1,fp);
             fread(buf,128,1,fp);
             fread(buf,128,1,fp);
   }

   /* Data decode loop:
      reads byte, tests to see if negative,
      if negative, then use as count for RLE
      of next byte in file; else just read
      that number of bytes. */

   while(fread(&cnt,1,1,fp)){
      if(cnt & 0x80){
         fread(&md,1,1,fp);
         cnt = (~cnt)+2;
         for(i=0; i<cnt; ++i){
            tmp = md;
            for(j=0; j<8; ++j){
               if(tmp&0x80)*imd = 0xff;
                  else *imd = 0;
               ++imd;
               tmp = (tmp<<1)&0xff;
            }
         }
      } else {
         cnt += 1;
         for(i=0; i<cnt; ++i){
            fread(&md,1,1,fp);
            for(j=0; j<8; ++j){
               if(md&0x80)*imd = 0xff;
                  else *imd = 0;
               ++imd;
               md = (md<<1)&0xff;
            }
         }
      }
   }

   fclose(fp);
}
```

SEE ALSO: Run Length Encoding

CLASS: Image Fundamentals

DESCRIPTION:

A *mask* defines a small image that is used to set parameters and define the area of operation to take place on a larger image when performing a *discrete convolution* or *correlation.* In the case of convolution, the mask is called the *filter* and when used in correlation, the mask is generally called a *template. Kernel* is also a term used to refer to a mask and is used primarily when discussing image transforms.

EXAMPLE:

$$\begin{bmatrix} -2 & 0 & +1 \\ -1 & 0 & +1 \\ -2 & 0 & +2 \end{bmatrix} \quad \begin{bmatrix} -2 & -1 & -2 \\ 0 & 0 & 0 \\ +2 & +1 & +2 \end{bmatrix}$$

4 × 4 Vertical and Horizontal Edge Detection Masks.

ALGORITHM:

```
/* Mask declaration and initialization
   for Vertical Edge Detector */

int Mask[3][3];

Mask[0][0] = -2;  Mask[0][1]=0;  Mask[0][2]=1;
Mask[1][0] = -1;  Mask[1][1]=0;  Mask[1][2]=1;
Mask[2][0] = -2;  Mask[2][1]=0;  Mask[2][2]=2;
```

The mask may also be described as a small image structure. This is the format used by the convolution and correlation algorithms described in this book. An example of this form is shown in Appendix B.

SEE ALSO: Discrete Convolution, Discrete Correlation, Spatial Filtering, Spatial Masks

CLASS: Mensuration

DESCRIPTION:

The *Maximum Axis* of an object is formally defined as the axis of minimum inertia (dispersion) passing through the *centroid*. There are a number of ways to calculate this, the most accurate being to compute the eigenvalues and eigenvectors of the scatter matrix comprised of the coordinate points of the object. The eigenvector corresponding to the largest eigenvalue will be the maximal axis. Another method is to fit an ellipse to the object perimeter to derive the maximum axis. A far simpler method uses central moments in the following formula to derive the slope, θ, of the Maximum Axis:

$$\tan^2\theta + \frac{\overline{m}_{20} - \overline{m}_{02}}{\overline{m}_{11}} \tan\theta - 1 = 0.$$

This quadratic can then be solved for tan θ.

The Maximum Axis is also known as the *Principal Axis* in the literature.

EXAMPLE:

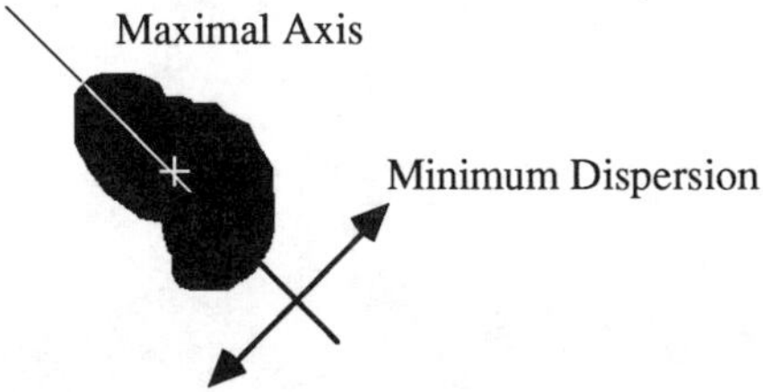

ALGORITHM:

The algorithm below assumes that the central moments have been calculated (**cm20, cm02, cm11**) and it returns the slope of the maximal axis. Note that the algorithm uses the include file **math.h** and the special functions **sqrt** and **atan**. The include file is a generalized standard amongst C compilers, as are the functions. If your library does not include these functions, **sqrt** returns the square root of its argument and **arctan** the arc tangent.

#include <math.h>

```
/* compute and return slope of maximum axis
*/
double max_axis_slope (struct Image *In,int
x1,int y1,int x2,int y2)
{
   float cm20,cm02,cm11,b;
   float theta;

   cm20 = cmoment(2,0,In,x1,y1,x2,y2);
   cm02 = cmoment(0,2,In,x1,y1,x2,y2);
   cm11 = cmoment(1,1,In,x1,y1,x2,y2);

   /* solve quadratic for tangent theta */
   b = (cm20 - cm02)/cm11;
   theta = ((-1.0*b)*sqrt(b*b + 4.0))/2.0;

   /* compute theta and return */
   return(atan(theta));
}
```

SEE ALSO: Minimum Axis, Centroid, Dilation, Moments

CLASS: Nonlinear Filters

DESCRIPTION:

The maximum filter is typically applied to an image to remove negative outlier noise. The maximum filter is also used in the computation of binary morphological dilation and is defined as

$$\text{Maximum}(A) = \max[\,A(x+i,\, y+j)\,]\,,$$

where the coordinate x + i, y + j is defined over the image A and the coordinate i, j is defined over the mask M. The mask M determines which pixels are used in the maximum calculation.

EXAMPLE:

(a)

(b)

(a) The original image and ***(b)*** the maximum filtered image using a 7 × 7 mask.

ALGORITHM:

The program assumes that the original image is a 256 graylevel × IMAGE->Rows × IMAGE->Cols pixel image stored in the structure IMAGE. The program performs an N × N maximum filter storing the resulting filtered image in the structure IMAGE1. The only restriction on the program is that the size of the mask N should be odd and be less than 12.

```
Maximum(struct Image *IMAGE, struct Image
*IMAGE1)
     {
     int X, Y, I, J, smax, N, a[11][11];
     N=3;
     for(Y=N/2; Y<IMAGE->Rows-N/2; Y++)
        {
        for(X=N/2; X<IMAGE->Cols-N/2; X++)
           {
           smax=0;
           for(J=-N/2; J<=N/2; J++)
             for(I=-N/2; I<=N/2; I++){
                a[I+N/2][J+N/2]= *(IMAGE->
                Data+X+I+(long)(Y+J)*
                IMAGE->Cols);}
           for(J=0; J<=N-1; J++)
              {
              for(I=0; I<=N-1; I++)
                 {
                 if(a[I][J] > smax)
                   smax = a[I][J];
                 }
              }
              *(IMAGE1->Data+X+(long)Y
              *IMAGE->Cols) =smax;
           }
        }
     }
```

SEE ALSO: Minimum, Arithmetic Mean, Median, and other Nonlinear Filters.

CLASS: Nonlinear Filters

DESCRIPTION:

A median filter operation on an image removes long tailed noise such as negative exponential and salt and pepper type noise from an image with a minimum blurring of the image. The median filter is defined as the median of all pixels within a local region of an image. Pixels that are included in the median calculation are specified by a mask. The median filter performs much better than the arithmetic mean filter in removing salt and pepper noise from an image and in preserving the spatial details contained within the image. The median filter can easily remove outlier noise from images that contain less than 50% of its pixels as outliers. The definition of a median filter in terms of an image A is

$$\text{Median}(A) = \text{Median}[A((x+i,\ y+j)] ,$$

where the coordinate x + i, y + j is defined over the image A and the coordinate i, j is defined over the mask M. The mask M determines which pixels are to be included in the median calculation.

EXAMPLE:

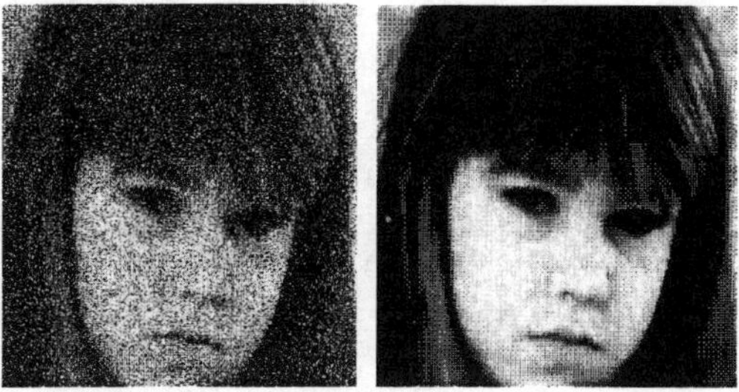

(a) *(b)*

(a) The original salt and pepper noise corrupted image and ***(b)*** the median filtered image using a 5×5 mask.

ALGORITHM:

The program assumes that the original image is a 256 graylevel × IMAGE->Rows × IMAGE->Cols pixel image stored in the structure IMAGE. The program then performs a N × N median filter on the image. The size of the filtering operation is determined by the variable N and should be set to an odd number and be less than 12. Upon completion of the program, the filtered image is stored in the structure IMAGE1.

```
Median(struct Image *IMAGE, struct Image
*IMAGE1)
      {
      int X, Y, I, J, Z;
      int N, AR[121], A;
      N=7;
      for(Y=N/2; Y<IMAGE->Rows-N/2; Y++)
         for(X=N/2; X<IMAGE->Cols-N/2; X++)
            {Z=0;
            for(J=-N/2; J<=N/2; J++)
               for(I=-N/2; I<=N/2; I++)
                  {
                  AR[Z]= *(IMAGE->Data+X
                  +I+(long)(Y+J)
                  *IMAGE->Cols);
                  Z++;
                  }
            for(J=1; J<=N*N-1;J++)
               {
               A = AR[J];
               I=J-1;
               while(I>=0 && AR[I] >A)
                  {
                  AR[I+1]=AR[I];
                  I=I-1;
                  }
                  AR[I+1]=A;
               }
               *(IMAGE1->Data+ X +(long)Y
               *IMAGE->Cols) = AR[N*N/2];
            }
      }
```

SEE ALSO: Arithmetic Mean, Minimum, Maximum, and other Nonlinear Filters

Mensuration Class

DESCRIPTION:

Mensuration is a sophisticated term for *measurement.* In image processing, it refers to the evaluation of features associated with objects extracted from images. These measurements may be used to classify objects, both for recognition schemes and for database purposes. Mensuration algorithms are well known to the medical imaging field for the evaluation of tumors, cell structures, and fetal tissues. Image object measurement is often tightly coupled to interactive graphic processing, where boundaries and objects are specified by a human operator using a pointing device. The algorithms presented here are useful in this instance or when objects are extracted automatically. It is beyond the scope of this book, however, to go into greater depth than a simple clustering algorithm for the automatic extraction of image objects.

Measures on the spatial distribution of pixels and pixel values are the most common and simplest to implement. These include the moments, area, and axes of image objects. There are a large number of measures that have been reported in the literature and only a sampling of them, the most popular, are presented in this book.

CLASS MEMBERSHIP:

- Area
- Centroid
- Circularity
- Clustering
- Compactness
- Maximum Axis
- Minimum Axis
- Moments
- Perimeter

CLASS: Nonlinear Filters

DESCRIPTION:

The midpoint filter is typically used to filter images containing short tailed noise such as Gaussian and uniform type noises. The midpoint filter output is the average of the maximum and minimum graylevel values within a local region of the image determined by a specified mask. The definition of the midpoint filter is

$$\text{Midpoint}(A) = \frac{\max[\,A(x+i,\,y+j)\,] + \min[\,A(x+i,\,y+j)\,]}{2},$$

where the coordinate x + i, y + j is defined over the image A and the coordinate i, j is defined over the mask M. The mask M determines which pixels are to be included in the range calculation.

EXAMPLE:

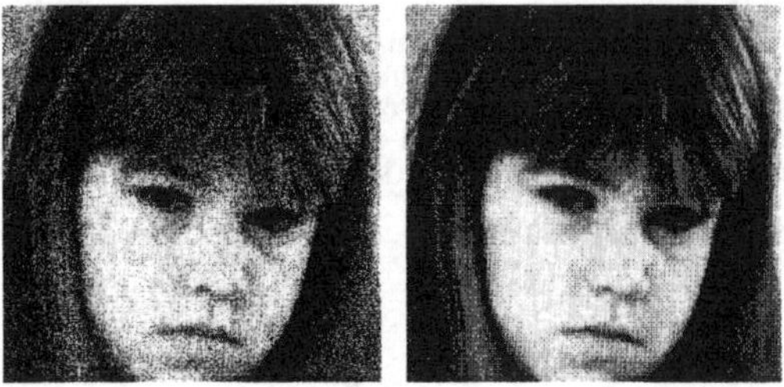

(a) *(b)*

(a) An uniform noise corrupted image (Variance = 800, Mean = 0) and ***(b)*** the midpoint filtered image using a 3 × 3 square mask.

ALGORITHM:

The program assumes that the original image is a 256 graylevel × IMAGE->Rows × IMAGE->Cols pixel image stored in the structure IMAGE. The program computes the midpoint filter over a set of pixels contained within a square N × N region of the image centered at the pixel X, Y. The size of the filtering operation is determined by the variable N and should be set to an odd number and be less than 12. Upon completion of the program, the midpoint filtered image is stored in the structure IMAGE1.

```
midpoint(struct Image *IMAGE, struct Image
*IMAGE1)
      {
      int X, Y, I, J, smin, smax, N;
      int a[11][11];
      N=3;
      for(Y=N/2; Y<IMAGE->Rows-N/2; Y++)
         for(X=N/2; X<IMAGE->Cols-N/2; X++)
            {
            smin=255; smax=0;
            for(J=-N/2; J<=N/2; J++)
              {
              for(I=-N/2; I<=N/2; I++)
                 {
                 a[I+N/2][J+N/2]=*(IMAGE->
                 Data+X+I+(long)(Y+J)
                 *IMAGE->Cols);
                 }
              }
            for(J=0; J<=N-1; J++)
               for(I=0; I<=N-1; I++)
                  {
                  if(a[I][J] < smin)
                    smin = a[I][J];
                  }
            for(J=0; J<=N-1; J++)
               for(I=0; I<=N-1; I++)
                  {
                  if(a[I][J] > smax)
                    smax = a[I][J];
                  }
            *(IMAGE1->Data+X+(long)Y
            *IMAGE->Cols)=(smax+smin)/2;
            }
      }
```

SEE ALSO: Geometric, Y_p, Harmonic, Arithmetic Mean, Median, and other Nonlinear Filters.

CLASS: Mensuration

DESCRIPTION:

Minimum Axis of an object is formally defined as the axis of maximum inertia (dispersion) passing through the *centroid*. There are a number of ways to calculate this, the most accurate being to compute the eigenvalues and eigenvectors of the scatter matrix comprised of the coordinate points of the object. The eigenvector corresponding to the smallest eigenvalue will be the minimal axis. Another method is to fit an ellipse to the object perimeter to derive the minimum axis.

If the maximum axis is known, then the minimum axis may be computed as the line 90° to it.

EXAMPLE:

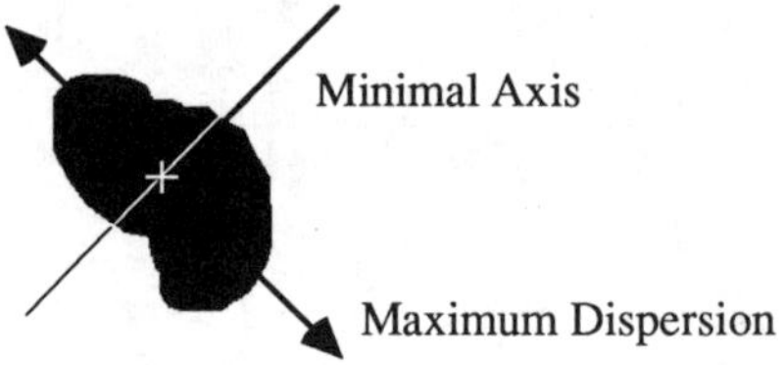

ALGORITHM:

The algorithm assumes that the central moments have been calculated (**cm20, cm02, cm11**) and returns the slope of the minimum axis as the line 90° to the maximum axis.

```
/* compute and return slope of minimum axis
*/
float min_axis_slope (struct Image *In,int
x1,int y1,int x2,int y2)
{
   return(max_axis_slope(In,x1,y1,x2,y2)+
                          (1.5714285714));
}
```

SEE ALSO: Maximum Axis, Centroid, Moments

CLASS: Nonlinear Filters

DESCRIPTION:

The minimum filter is typically applied to an image to remove positive outlier noise. The minimum filter is also used in the computation of binary morphological erosion and is defined as

$$\text{Minimum}(A) = \min[\,A(x+i,\, y+j)\,]\,,$$

where the coordinate x + i, y + j is defined over the image A and the coordinate i, j is defined over the mask M. The mask M determines which pixels are used in the minimum calculation.

EXAMPLE:

(a)

(b)

(a) The original image and *(b)* the minimum filtered image using a 7 × 7 mask.

ALGORITHM:

The program assumes that the original image is a 256 graylevel × IMAGE->Rows × IMAGE->Cols pixel image stored in the structure IMAGE. The program performs an N × N minimum filter storing the resulting filtered image in the structure IMAGE1. The only restriction on the program is that the size of the mask N should be odd and be less than 12.

```
Minimum(struct Image *IMAGE, struct Image
*IMAGE1)
    {
    int X, Y, I, J, smin, N, a[11][11];
    N=5;
    for(Y=N/2; Y<IMAGE->Rows-N/2; Y++)
        {
        for(X=N/2; X<IMAGE->Cols-N/2; X++)
            {
            smin=255;
            for(J=-N/2; J<=N/2; J++)
              for(I=-N/2; I<=N/2; I++)
                {
                a[I+N/2][J+N/2]= *(IMAGE->
                Data+X +I+(long)(Y+J)*IMAGE->
                Cols);
                }
            for(J=0; J<=N-1; J++)
                {
                for(I=0; I<=N-1; I++)
                    {
                    if(a[I][J] < smin)
                      smin = a[I][J];
                    }
                }
            *(IMAGE1->Data+X+(long)Y
            *IMAGE->Cols) = smin;
            }
        }
    }
```

SEE ALSO: Maximum, Arithmetic Mean, Median, and other Nonlinear Filters.

CLASS: Mensuration

DESCRIPTION:

Moments are a measure of the distribution of values across an axis. In imaging, we use two-dimensional moments to describe distributions of grayscale values. Moments are scalars and often provide excellent measures of an object for pattern analysis or recognition purposes. The two-dimensional moments, m_{pq}, are given by:

$$m_{pq} = \sum_{x=0}^{M-1} \sum_{y=0}^{N-1} x^p y^q \ f(x,y).$$

We may also define the central moments, given by:

$$\mu_{pq} = \sum_{x=0}^{M-1} \sum_{y=0}^{N-1} (x - \overline{x})^p (y - \overline{y})^q \ f(x,y),$$

where

$$\overline{x} = \frac{m_{10}}{m00}, \quad \overline{y} = \frac{m_{01}}{m00}.$$

EXAMPLE:

Xc = 39/13 = 3

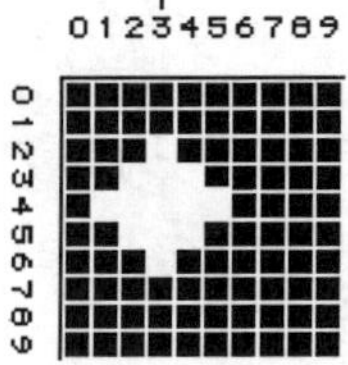

Yc = 52/13 = 4

The centroid of an object (above) is nothing more than the first central moment.

ALGORITHM:

The code given below calculates a desired central moment specified by **p** and **q** on a rectangular image region given by **x1**, **y1**, **x2**, **y2** in unsigned character image **In**. It returns the requested moment as the function's double value.

The macro **pix** is used for simplified access to the pixels of the image by coordinates.

```
#define pix(Im,x,y) \
        *(Im->Data + (x)*Im->Cols + (y))

/* Compute central moments */

double cmoment(int p,int q,
               struct Image *In,
               int x1,int y1,int x2,int y2)
{
   int i,j,xb,yb;
   double  m00=0.0,m10=0.0,m01=0.0,upq=0.0;

   /* code takes advantage of a number of
      symmetry properties of the central
      moments */

   if((p==1 && q==0) || (p==0 && q==1))
      return(upq);

   for(i=x1; i<x2; ++i)
      for(j=y1; j<y2; ++j)
         m00 += pix(In,i,j);

   if(p==0 && q==0)return(m00);

   for(i=x1; i<x2; ++i)
      for(j=y1; j<y2; ++j)
         m10 += j * (pix(In,i,j));

   for(i=x1; i<x2; ++i)
      for(j=y1; j<y2; ++j)
         m01 += i * (pix(In,i,j));

   xb = floor(0.5 + (m10/m00));

   yb = floor(0.5 + (m01/m00));
```

```
    if(p == 0){
       for(i=x1; i<x2; ++i)
          for(j=y1; j<y2; ++j)
             upq += pow(j-yb,q)*pix(In,i,j);
       return(upq);
    }

    if(q == 0){
       for(i=x1; i<x2; ++i)
          for(j=y1; j<y2; ++j)
             upq += pow(i-yb,p)*pix(In,i,j);
       return(upq);
    }

    for(i=x1; i<x2; ++i)
       for(j=y1; j<y2; ++j)
          upq += pow(i-xb,p)*
                     pow(j-yb,q)*pix(In,i,j);

    return(upq);
}
```

SEE ALSO: Centroid

CLASS: Graphics Algorithms

DESCRIPTION:

Morphing is used to distort one image into another, or provide a smooth transition from one image to another and thus create the illusion of a transformation. The term was coined by the special effects industry and implies using an image processing algorithm to effect a dynamic shape change from one object into another. One may define a simple morphing that uses a fade or dissolve process that decrements or increments pixel values from a source image until they equal pixel values from a destination image. More advanced morphing uses a warping process in addition to the fade/dissolve to slowly bring into alignment the shapes of the start and end pictures. This adds to the smooth transition of the process and yields a dramatic effect when "morphing" one human face into another or a human face into (or from) an inanimate object or animal.

The example shows how objects of the same shape can be "morphed" using a fade from the start image to the end image. In this case, the Panic key is changed into the Delete key.

EXAMPLE:

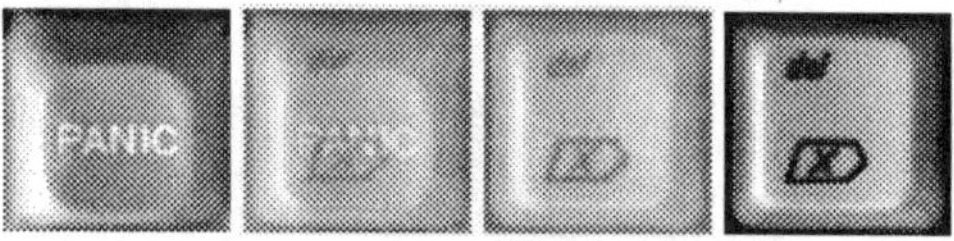

ALGORITHM:

The algorithm given fades one image into another and is intended to be called until completion. Three images are passed to the routine with a completion (**Done**) and status (**Stat**) flag. The first two image structure pointers are the **Start** and **End** images, and the third is the working image, **Out**, that is displayed after each call to the routine. The flags must be initialized to zero on the first call and all images passed must be of congruent dimension. When the routine is finished, the **Done** flag will be nonzero.

When called initially, **Stat** and **Done** equal zero, the **Start** image is copied to the **Out** and the routine returns. Successive calls increment or decrement the **Out** image pixels until they equal the **End** image. This will occur after no more than 256 calls. A faster fade can be achieved by modifying the increment-decrement loop.

A sequence of fades can be saved and warping done on the images in the sequence to effect a more sophisticated morph.

```
/* Simple fading Morph routine */
morph_fade(struct Image *Start,
           struct Image *End,
           struct Image *Out,
           unsigned char *Stat, char *Done)
{
   long i,j,sz;
   unsigned char *tmpO,*tmpE;

   if(*Done)return;
   tmpO = Out->Data;

   if(*Stat == 0){
      for(i=0;i<Start->Rows*Start->Cols;++i)
         *(tmpO++) = *(Start->Data + i);
      *Stat = 1;
      return;
   }
   ++(*Stat);
   if(*Stat==256){ *Done = 1; return;}

   tmpE = End->Data;
   sz = Start->Rows*Start->Cols;

   /* Increment/Decrement Loop */
   for(i=0;i<sz;++i,++tmpO,++tmpE){
      if(*tmpO == *tmpE)continue;
      if(*tmpO < *tmpE)*tmpO += 1;
         else *tmpO -= 1;
   }
}
```

SEE ALSO: Warping

Morphological Filters

DESCRIPTION:

The image processing of an image prior to pattern recognition and identification usually involves the morphological filtering of an image to change the geometrical shape of objects within an image. The goal of morphological filters is to smooth the contours of objects and to decompose an image into its fundamental geometrical shapes for image recognition.

Morphological filtering operations can be separated into two categories: (1) binary morphological filtering of binary images, and (2) graylevel morphological filtering of graylevel images. All of the filtering operations for both binary and graylevel morphological filtering are derived from the two fundamental morphological operations of *dilation* and *erosion.*

CLASS MEMBERSHIP:

Binary:

- Closing Filter
- Dilation Filter
- Erosion Filter
- Hit-Miss Filter
- Opening Filter
- Outline Filter
- Skeleton Filter
- Thickening Filter
- Thinning Filter

Graylevel:

- Closing Filter
- Dilation Filter
- Erosion Filter
- Opening Filter
- Top-hat Filter

CLASS: Segmentation

DESCRIPTION:

Multi-graylevel thresholding is used in image processing to separate different graylevel regions within an image. Multi-graylevel thresholding converts a multi-graylevel image into an image containing a small number of graylevel values. The multi-graylevel threshold operation defined over four graylevel regions is

$$g(x, y) = \begin{cases} G_3 & \text{if } f(x, y) > T_3 \\ G_2 & \text{if } T_2 < f(x, y) \le T_3 \\ G_1 & \text{if } T_1 < f(x, y) \le T_2 \\ G_0 & \text{if } f(x, y) \le T_1 \end{cases},$$

where f(x, y) is the original image, g(x, y) is the multi-graylevel thresholded image, T_1, T_2, and T_3 are the threshold values, and G_0, G_1, G_2, and G_3 are the graylevel values used in the multi-graylevel thresholded image.

EXAMPLE:

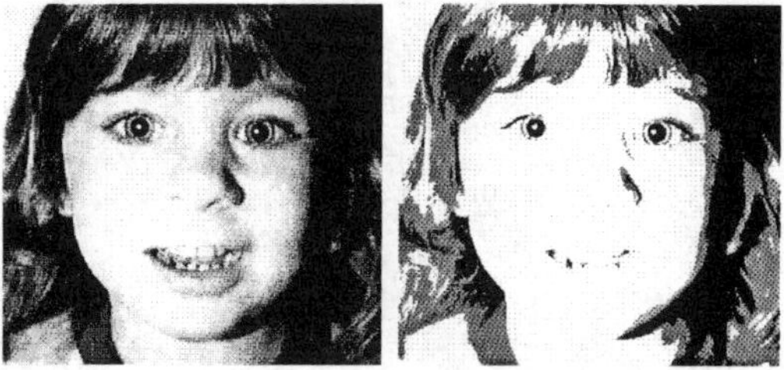

(a) *(b)*

(a) The original image and *(b)* the thresholded image using two thresholds of $T_1 = 40$ and $T_2 = 113$ and three graylevel values of $G_0 = 0$, $G_1 = 128$, and $G_2 = 255$.

ALGORITHM:

The program assumes that the original image is a 256 graylevel × IMAGE->Rows × IMAGE->Cols pixel image stored in the structure IMAGE. The program computes a 3-graylevel image by comparing each pixel of the image against the threshold parameters THRES1 and THRES2. If a pixel's graylevel value is greater than the THRES1 parameter, it is set to the graylevel value given by the G2 parameter and if the pixel's graylevel value is between THRES1 and THRES2 the pixel's graylevel value is set to the graylevel value given by the G1 parameter; otherwise it is set to the graylevel value given by the G0 parameter. The resulting threshold image is returned by the program in the structure IMAGE1.

```
MultiThreshold(struct Image *IMAGE, struct
Image *IMAGE1, int THRES1, int THRES2)
      {
      int X, Y, G2, G1, G0, GR;
      G2=255;
      G1=128;
      G0=0;
      for(Y=0; Y<IMAGE->Rows; Y++)
         {
         for(X=0; X<IMAGE->Cols; X++)
            {
            GR=*(IMAGE->Data+X+(long)Y*
            IMAGE->Cols);
            if(GR > THRES2)
                  GR = G2;
            if(GR <= THRES2 &&
                  GR > THRES1)
                  GR= G1;
            if(GR < THRES1)
                  GR= G0;
            *(IMAGE1->Data+X+(long)Y*
            IMAGE->Cols)= (unsigned char)GR;
            }
         }
      }
```

SEE ALSO: Thresholding, Point Detector, Line Detectors, and Optimum Thresholding

CLASS: Noise

DESCRIPTION:

Negative exponential type noise is the result of acquiring an image illuminated with a coherent laser. The optics community refers to this type of noise as *laser speckle.* Its histogram is defined as

$$h_i = \frac{\exp^{-(G_i/a)}}{a} \qquad \text{for } 0 \le G_i < \infty \ ,$$

where G_i is the ith graylevel value of the image and a^2 is the variance.

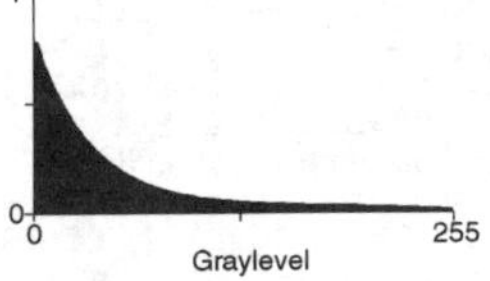

A histogram of negative exponential noise.

EXAMPLES:

(a)

(b)

(a) The original image and ***(b)*** the negative exponential noise degraded image with a variance = 800.

ALGORITHM:

The program generates a negative exponential noise image of 256 graylevel × IMAGE->Rows × IMAGE->Cols pixel image stored in the structure IMAGE. The program assumes that the function rand() generates a uniform random number in the range of 0 to 32767. The desired variance is passed to the program upon execution. If the noise graylevel value generated exceeds the 256 graylevel range, the noise graylevel value is truncated to either 0 or 255.

```
NegExp(struct Image *IMAGE,
      float VAR)
      {
      int X, Y;
      float NOISE, A, theta, Rx, Ry, Rz;
      for(Y=0; Y<IMAGE->Rows; Y++)
         for(X=0; X<IMAGE->Cols; X++)
            {
            A= sqrt((double)VAR)/2;
            NOISE=sqrt(-2 * A * log(1.0-
            (float)rand() / 32767.1));
            theta=(float)rand()*
            1.9175345E-4 - 3.14159265;
            Rx = NOISE*cos(theta);
            Ry = NOISE*sin(theta);
            NOISE = Rx*Rx + Ry*Ry;
            if(NOISE > 255)
                  NOISE = 255;
            *(IMAGE->Data+X+(long)Y*
            IMAGE->Cols) = (unsigned
            char)(NOISE +.5);
            }
      }
```

SEE ALSO: Gaussian, Uniform, Salt and Pepper, Rayleigh and Gamma Noises

Noise

DESCRIPTION:

The ability to add a controlled amount of noise to an image is imperative in the development of image processing filtering algorithms. The controlled degradation of noise free images allows for an easy comparison of a filter's performance against various noise conditions that are encountered. For example, *Gaussian* type noise is better filtered using a mean filter, while *Salt and Pepper* type noise is better filtered using a median filter. *Gaussian* noise appearing in an image is usually the result of camera electronic noise, while *Salt and Pepper* noise is the result of inoperative or "dead" pixels within the camera sensor.

CLASS MEMBERSHIP:

Gamma Noise

Gaussian Noise

Negative Exponential Noise

Rayleigh Noise

Salt and Pepper Noise

Uniform Noise

SEE ALSO: Spatial, Nonlinear and Adaptive Filters

Nonlinear Filters

DESCRIPTION:

A large portion of image processing filters fall under the nonlinear filtering category. Nonlinear filters operate on an image by computing a given nonlinear function over a local window and replacing a specified pixel within the local window with this value. One of the most important of the nonlinear filters which is based upon order statistics is the median filter. It is typically used to remove *salt and pepper* noise from an image while offering the advantages over the arithmetic mean filter of preserving the edge information within an image. Nonlinear filters are not just limited to the removal of noise from an image. The range filter, for example, can be used to detect edges within an image.

CLASS MEMBERSHIP:

Alpha-Trimmed Mean Filter

Contra-Harmonic Filter

Geometric Mean Filter

Harmonic Mean Filter

Maximum Filter

Median Filter

Midpoint Filter

Minimum Filter

Range Filter

Weighted Median Filter

Yp Mean Filter

SEE ALSO: Spatial and Adaptive Filters

CLASS: Histogram Operation

DESCRIPTION:

Nonlinear graylevel transformation of an image is used to increase the contrast for certain graylevel values of an image while decreasing the contrast for other graylevel values. Graylevel transformation **T** of an image is defined as

$$s_i = \mathbf{T}(g_i) \ ,$$

where g_i is the ith graylevel value of the original image and s_i is the ith graylevel value of the nonlinear graylevel transformed image.

EXAMPLE:

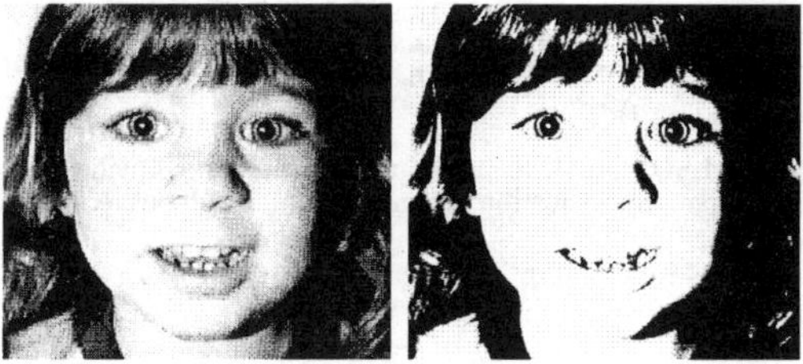

(a) *(b)*

(a) The original image and ***(b)*** the non-linear graylevel transformed image.

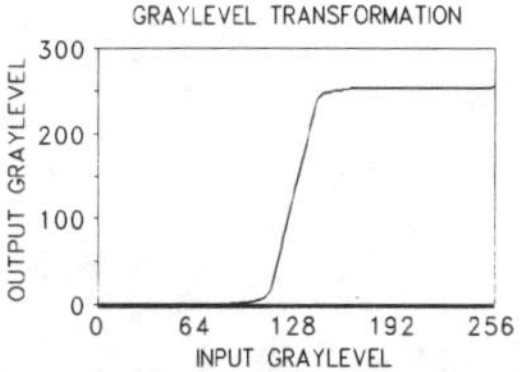

The specified nonlinear transformation.

ALGORITHM:

The program assumes that the original image is a 256 graylevel × IMAGE->Rows × IMAGE->Cols pixel image stored in the structure IMAGE and that the desired graylevel transformation is stored in the integer array TRANS[]. The program then transforms the original image's graylevels using the TRANS[] array. Upon completion of the program, the new image is stored in the structure IMAGE1.

```
NonLinGraylevel(struct Image *IMAGE, struct
Image *IMAGE1, int TRANS[])
      {
      int X,Y;
      for(Y=0; Y<IMAGE->Rows; Y++)
         for(X=0; X<IMAGE->Cols; X++)
            *(IMAGE1->Data+X+(long)Y*
            IMAGE->Cols)=
            TRANS[*(IMAGE->Data+X+(long)Y*
            IMAGE->Cols)];
      }
```

SEE ALSO: Graylevel Histogram, Histogram Equalization, Histogram Specification, Contrast, and Brightness Correction

Opening (Binary) Filter

CLASS: Morphological Filters

DESCRIPTION:

Morphological opening of a binary object is defined as the erosion of that object followed by the dilation of the eroded object. The opening filter operation will reduce small outward bumps and small narrow openings.

$$\text{Open}(A, B) = (A \ominus B) \oplus B$$

EXAMPLES:

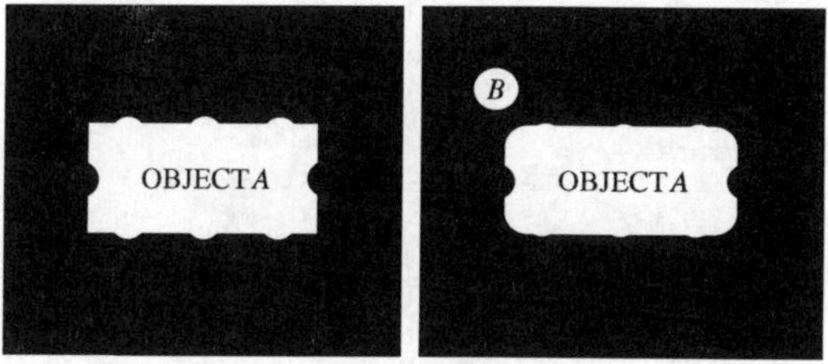

(a) ***(b)***

(a) The original binary image of object A and ***(b)*** the opened image of object *A* with a circular structuring function *B*.

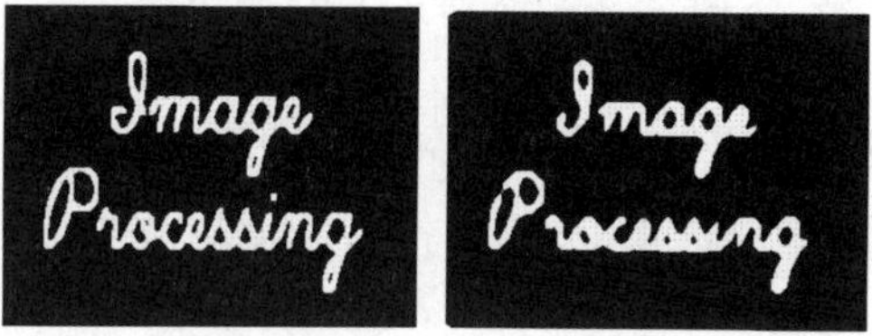

(a) ***(b)***

(a) The original image and ***(b)*** the opened image.

ALGORITHM:

The program assumes that the original binary image is an IMAGE->Rows × IMAGE->Cols pixel image with a background graylevel value of 0 and an object graylevel value of 255 (object) and is stored in the structure IMAGE. The N × N structuring function is stored in array MASK[][]. Upon completion of the program, the opened image is stored in the structure FILTER. The binary erosion and dilation functions used by the algorithm can be found under binary erosion and dilation respectively.

```
#define N 5

Open(struct Image *IMAGE, int
  MASK[][N], struct Image *FILTER)
    {
    int X, Y;
    Erosion(IMAGE, MASK, FILTER);
    for(Y=0; Y<IMAGE->Rows; Y++)
        {
        for(X=0; X<IMAGE->Cols; X++)
            {
            *(IMAGE->Data+ X +
            (long)Y * IMAGE->Cols) =
            *(FILTER->Data+ X + (long)Y
            *IMAGE->Cols);
            }
        }
    Dilation(IMAGE, MASK, FILTER);
    FILTER->Rows=IMAGE->Rows;
    FILTER->Cols=IMAGE->Cols;
    }
```

SEE ALSO: Binary Dilation, Erosion, and Closing Filters

CLASS: Morphological Filters

DESCRIPTION:

Morphological opening of an image is defined as the graylevel erosion of an image followed by the graylevel dilation of the eroded image. The opening filter operation will reduce small positive oriented graylevel regions and positive graylevel noise regions generated from salt and pepper noise.

$$\mathrm{Open}(A, B) = (A \ominus B) \oplus B$$

EXAMPLES:

(a)

(b)

(a) The original image and ***(b)*** the opened image using an all zero 7×7 mask.

ALGORITHM:

The program assumes that the original image is a 256 graylevel × IMAGE->Rows × IMAGE->Cols pixel image stored in the structure IMAGE. The N × N structuring function is stored in the array MASK[][]. Upon completion of the program, the opened image is stored in the structure FILTER. The graylevel erosion and dilation functions used by the algorithm can be found under graylevel erosion and dilation respectively.

```
#define N 5

OpenGray(struct Image *IMAGE, int
   MASK[][N],struct Image *FILTER)
      {
      int X, Y;
      ErosionGray(IMAGE, MASK, FILTER);
      for(Y=0; Y<IMAGE->Rows; Y++)
         {
         for(X=0; X<IMAGE->Cols; X++)
            {
            *(IMAGE->Data+X+(long)Y*
            IMAGE->Cols)= *(FILTER->
            Data+X+(long)Y*IMAGE->Cols);
            }
         }
      DilationGray(IMAGE, MASK, FILTER);
      FILTER->Rows=IMAGE->Rows;
      FILTER->Cols=IMAGE->Cols;
      }
```

SEE ALSO: Graylevel Dilation, Erosion, Top-Hat, and Closing Filters

CLASS: Segmentation

DESCRIPTION:

Optimum thresholding of an image assumes that an image's histogram contains two predominant peaks, one due to the background and one due to an object. The goal of optimum thresholding is to find the valley between the two peaks and threshold the image at this graylevel value. The histogram below illustrates the location of the optimum threshold value.

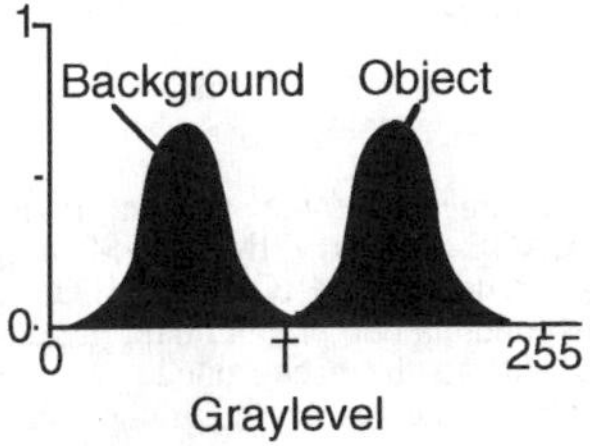

EXAMPLE:

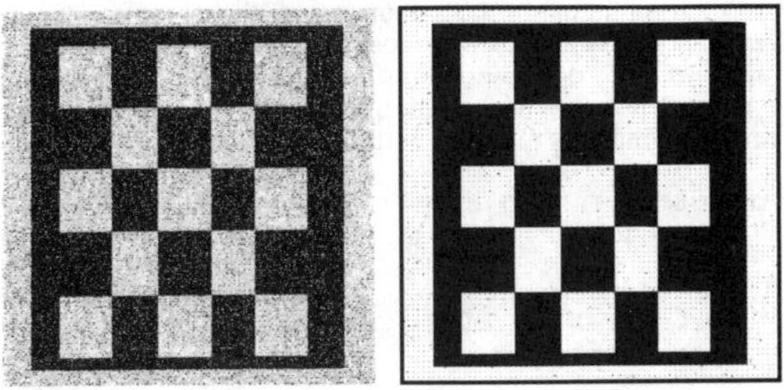

(a) *(b)*

(a) The original image containing additive Gaussian noise containing a variance of 600 and ***(b)*** the optimum thresholded image with T = 127 .

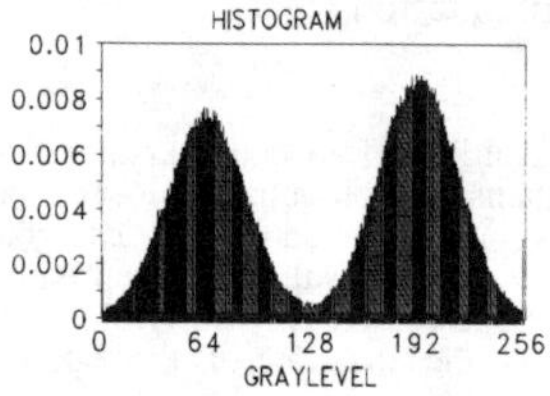

The histogram of the original image showing two predominant peaks.

ALGORITHM:

The program assumes that the original image is a 256 graylevel × IMAGE->Rows × IMAGE->Cols pixel image stored in the structure IMAGE and that the image's histogram only contains two predominant peaks, one due the object and one due to the background. The program first starts by computing the histogram of the image using the algorithm given in this text. The program then smoothes the histogram curve using a 31 × 1 mean filter to reduce error in finding the optimum threshold value. Finally, ignoring the two end points of the histogram curve, the program locates the valley between the two peaks and uses this graylevel value as the optimum threshold value in creating the binarized image. Upon completion of the program, the binarized image is returned in the structure IMAGE1.

```
OptimumThreshold(struct Image *IMAGE,
struct Image *IMAGE1)
    {
    int X, Y, FLAG, J, THRES, GO, GB;
    float HIST[256], SUM;
    GO=255;
    GB=0;
    Histogram(IMAGE, HIST);
    for(Y=0;Y<=255;Y++)
        {
        SUM=0.0;J=0;
        for(X=-15; X<=15; X++)
            {
```

```
            J++;
            if((Y-X)>=0)
               SUM= SUM + HIST[Y-X];
            }
         HIST[Y]=SUM/(float)J;
       }
     Y=2;
     FLAG=0;
     THRES=0;
     while(FLAG==0 && Y< 254)
         {
         if((HIST[Y-1] >= HIST[Y]) &&
         (HIST[Y] < HIST[Y+1]))
           {
           FLAG=1;
           THRES=Y;
           }
         Y++;
         }
     for(Y=0; Y<=511; Y++)
        {
        for(X=0; X<=511; X++)
           {
           if(*(IMAGE->Data+X+(long)Y*
           IMAGE->Cols) > THRES)
                 *(IMAGE1->Data+X+(long)Y*
                 IMAGE->Cols)= GO;
           else
                 *(IMAGE1->Data+X+(long)Y*
                 IMAGE->Cols)= GB;
           }
        }
     }
```

SEE ALSO: Multi-graylevel Thresholding, Point Detector, Line Detector, and Thresholding

CLASS: Morphological Filters

DESCRIPTION:

Binary outlining of an object is the process of changing all of an object's pixels to the background graylevel value except those pixels that lie on the object's contour. Binary outlining is defined as the eroded image subtracted from the original image or the original image subtracted from the dilated image. The width of the contour is determined by the size of the structuring function B.

$$\text{Outline}(A, B) = A - (A \ominus B) \ ,$$

or

$$\text{Outline}(A, B) = (A \oplus B) - A$$

EXAMPLES:

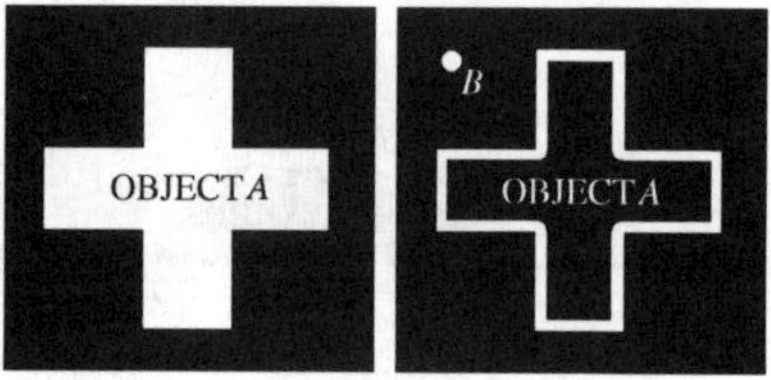

(a) ***(b)***

(a) The original binary image and ***(b)*** the outlined image.

(a) ***(b)***

(a) The original image and ***(b)*** the outlined image.

ALGORITHM:

The program assumes that the original binary image is an IMAGE->Rows × IMAGE->Cols pixel image with a background graylevel value of 0 and an object graylevel value of 255 (object) and is stored in the structure IMAGE. The N × N structuring function is stored in array MASK[][]. Upon completion of the program, the outlined image is stored in the structure FILTER. This program is modified to the second binary outline filter by changing the following lines:

(Old) *(FILTER->Data + X + (long)Y*IMAGE->Cols) =
(IMAGE->Data + X + (long)Y IMAGE->Cols) -
(FILTER->Data + X + (long)Y IMAGE->Cols);

(New) *(FILTER->Data + X + (long)Y*IMAGE->Cols) =
(FILTER->Data + X + (long)Y IMAGE->Cols) -
(IMAGE->Data + X + (long)Y IMAGE->Cols);

(Old) Erosion(IMAGE, MASK, Filter);
(New) Dilation(IMAGE, MASK, Filter);

```
#define N 5

Outline(struct Image *IMAGE,int MASK[][N],
      struct Image *FILTER)
      {
      int X,Y, I, J;
      Erosion(IMAGE, MASK, FILTER);
      for(Y=0; Y<IMAGE->Rows; Y++)
         for(X=0; X<IMAGE->Cols; X++)
         *(FILTER->Data + X + (long)Y*
         IMAGE->Cols)= *(IMAGE->Data + X
         + (long)Y* IMAGE->Cols) -
         *(FILTER->Data + X + (long)Y*
         IMAGE->Cols);
      FILTER->Rows=IMAGE->Rows;
      FILTER->Cols=IMAGE->Cols;
      }
```

SEE ALSO: Edge Detection, Binary Erosion, Dilation, Closing, and Opening Filters

CLASS: Storage Formats

DESCRIPTION:

PC Paintbrush, or PCX, files are images created or intended to be accessed by the PC Paintbrush program by ZSoft. This sofware is classified as a drawing program and allows the user to draw and fill objects, add text to pictures, and manipulate color or grayscale palettes.

ALGORITHM:

PCX files consist of a 128-byte header followed by data. The header for these files may be described by the following structure:

```
struct PCX_header{
   char manufacturer; /* always 0x0a */
   char version;
   char encoding;      /* always 1 */
   char pixel_depth;
   int  x_min,y_min;  /* image origin */
   int  x_max,y_max;  /* image size */
   int  hor_res;       /* horiz/vert. */
   int  ver_res;       /* resolution  */
   char palette[48];  /* color palette for
                          16-color images */
   char reserved;
   char color_planes; /* # of planes */
   int  line_bytes;    /* line buffer size */
   int  palette_type; /* 1: grayscale
                          2: color   */
   char filler[58];    /* fill to make header
                          128-bytes total */
};
```

Bytes following the header are run length encoded image data compressed line-by-line. The length of an uncompressed line is given by the *line_bytes* variable from the header, hence the decompression algorithm knows when to stop decompressing a line. The byte-by-byte decoding starts by examining the two most significant bits of the compressed byte. If these are set, the byte is accepted as is. If not, the byte becomes a run length count when ANDed with 00111111_2 (or $3F_{16}$) and the byte following is duplicated accordingly. Decompression stops when *line_bytes* have been extracted from the compressed data.

Two routines are provided, the first, **rd_PCX_hdr**, reads in and returns a PCX header structure pointer, **rt_pcx**, from the binary file stream pointed to by **fp**. A zero is returned if **fp** does not contain a PCX file. Note also the use of the **rword** routine, which is described under the TIF topic description. This routine may or may not be needed depending on the target system being used. If erroneous results are returned from a known PCX file, then the byte order is not swapped and the last value passed to **rword** should be changed to zero.

The second routine, **decode_PCX**, should be called after **rd_PCX_hdr**. The routine decodes the RLE PCX data from the file stream **fp** into the data buffer of the image structure pointed to by **In**. The **Rows** and **Cols** variables of **In** must be initialized from the PCX header data. The decoder assumes an 8 bits per pixel image.

```
rd_PCX_hdr(FILE *fp,
           struct PCX_header *rt_pcx)
{
   fread(&rt_pcx->manufacturer,1,1,fp);
   if(rt_pcx->manufacturer!=0x0a)return(0);

   fread(&rt_pcx->version,1,1,fp);
   fread(&rt_pcx->encoding,1,1,fp);
   fread(&rt_pcx->pixel_depth,1,1,fp);

   rt_pcx->x_min    =  rword(fp,1);
   rt_pcx->y_min    =  rword(fp,1);
   rt_pcx->x_max    =  rword(fp,1);
   rt_pcx->y_max    =  rword(fp,1);
   rt_pcx->hor_res  =  rword(fp,1);
   rt_pcx->ver_res  =  rword(fp,1);
   rt_pcx->ver_res  =  rword(fp,1);
   rt_pcx->ver_res  =  rword(fp,1);

   fread(&rt_pcx->palette,48,1,fp);
   fread(&rt_pcx->reserved,1,1,fp);
   fread(&rt_pcx->color_planes,1,1,fp);

   rt_pcx->line_bytes   = rword(fp,1);
   rt_pcx->palette_type = rword(fp,1);
   fclose(fp);
   return(1);
}
```

```
decode_PCX(FILE *fp,struct Image *In)
{
  long i;
  unsigned char *ind, c, j, tmp[128];

  /* skip PCX header */
  fread(tmp,128,1,fp);
  ind = In->Data;

  while(i<(long)In->Rows * (long)In->Cols){
     c = (unsigned char)(fgetc(fp)&0xff);
     if((c&0xc0) == 0xc0){
        j = c & 0x3f;
        c = (unsigned char)(fgetc(fp)&0xff);
        while(j--){
              *ind++ = c;
              ++i;
        }
     } else {
        *ind++ = c;
        ++i;
      }
  }

   fclose(fp);
}
```

SEE ALSO: Run Length Encoding, Tagged Interchange File Format (TIF)

Perimeter

CLASS: Mensuration

DESCRIPTION:

Perimeter is the number of pixels in an object's boundary. The boundary may be computed by subtracting the erosion of an object from the original object, or by using an edge detector to trace the boundary. The area function can then be used to compute the number of pixels left in the boundary, hence the perimeter. The perimeter is useful in computing the compactness of an object.

No algorithm is given here, however, the sequence of steps used to compute the perimeter of an object in a region is shown in the example.

EXAMPLE:

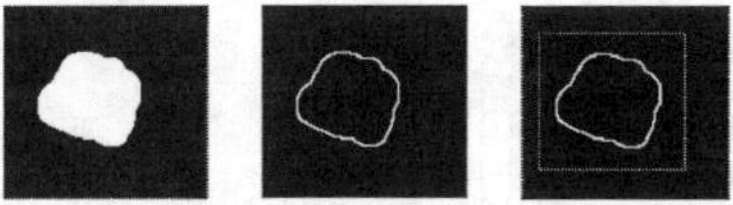

Left: Original object.

Middle: Object boundary (single pixel erosion).

Right: Region for area calculation.

SEE ALSO: Area, Compactness, Erosion, Edge Detection

CLASS: Image Fundamentals

DESCRIPTION:

A *pixel*, or Picture Element, is the smallest unit possible in an image. The physical size of a pixel is determined by the display or output device that expresses it, while the computational size is limited only by the memory of the machine processing the picture. The *sampling* of an image is an expression of pixel size. The depth of the pixel expresses the level of *quantization*. An image where pixels are only one bit deep is a binary image. The example below shows a *graylevel* image with two levels of expansion to show the individual pixels.

EXAMPLE:

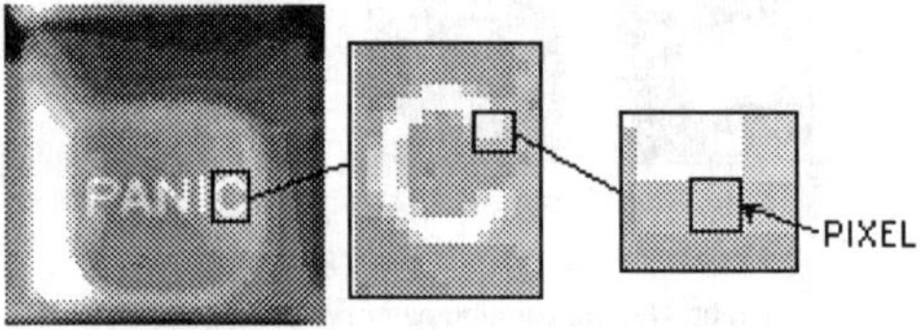

PROGRAM EXAMPLE:

```
/* 512 x 512 8-bit/pixel */
   char Image[512][512];
/* 256 x 256 floating point image, pixel
   range determined by float definition  */
   float RealImage[256][256];
/* 512 x 512 Image Structure */
   struct Image In;
   In.Rows = In.Cols = 512;
   /* In[i][j] = 255; */
   *(In.Data + i*In.Cols + j) = 255;
```

SEE ALSO: Quantization, Sampling

CLASS: Segmentation

DESCRIPTION:

Point detection can be used to find point discontinuities within an image. Point detection within an image is accomplished by performing spatial filtering using the following 3×3 mask

-1	-1	-1
-1	8	-1
-1	-1	-1

The 3×3 spatial filtering operation reduces to

$$\begin{aligned} g(x, y) = 8 \cdot f(x, y) &- f(x-1, y-1) - f(x-1, y) \\ &- f(x-1, y+1) - f(x, y-1) - f(x, y+1) \\ &- f(x+1, y-1) - f(x+1, y) \\ &- f(x+1, y+1) \end{aligned}$$

where f(x, y) is the original image and g(x, y) is the point detected image.

EXAMPLE:

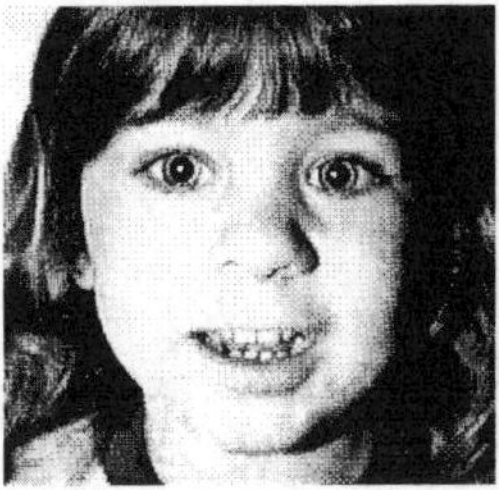

(a)

(b)

(a) The original image and *(b)* the point detected image using the 3×3 mask given in the text.

ALGORITHM:

The program assumes that the orignal image is a 256 graylevel × IMAGE->Rows × IMAGE->Cols pixel image stored in the structure IMAGE. The program then computes the point detected image using the 3 × 3 mask given in the text. Upon completion of the program, the line detected image is stored in the structure IMAGE1.

```
PointDetector(struct Image *IMAGE, struct
Image *IMAGE1)
      {
      int X, Y, I, J, SUM;
      int MASK[3][3];
      MASK[0][0]=-1;MASK[0][1]=-1;
      MASK[0][2]=-1;MASK[1][0]=-1;
      MASK[1][1]= 8;MASK[1][2]=-1;
      MASK[2][0]=-1;MASK[2][1]=-1;
      MASK[2][2]=-1;
      for(Y=1; Y<IMAGE->Rows-1; Y++)
         {
         for(X=1; X<IMAGE->Cols-1; X++)
            {
            SUM=0;
            for(I=-1;I<=1;I++)
               for(J=-1;J<=1;J++)
                  {
                  SUM=SUM+
                  *(IMAGE->Data+X+I+
                  (long)(Y+J)*IMAGE->Cols)*
                  MASK[I+1][J+1];
                  }
            if(SUM>255)
               SUM=255;
            if(SUM<0)
               SUM=0;
            *(IMAGE1->Data+X+(long)Y*
            IMAGE->Cols)=SUM;
            }
         }
      }
```

SEE ALSO: Thresholding, Multi-graylevel Thresholding, Line Detectors, and Optimum Thresholding

CLASS: Color Image Processing

DESCRIPTION:

To pseudocolor a graylevel image, the individual graylevels within the image must be mapped to a set of red, blue, and green color images.

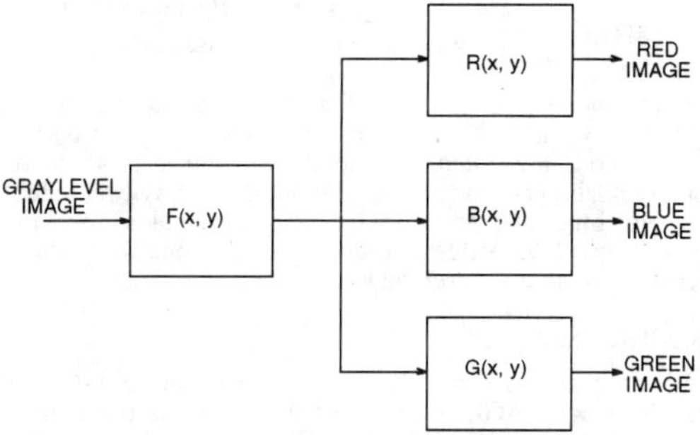

(a) A block diagram that implements pseudocoloring.

Consider an image with N discrete graylevel values represented by the function F. An image can be considered as a collection of graylevel values given by the function F for all the pixels within the image. Next, this function, F, can be mapped to three functions R, B, and G that produce the output colors red, blue, and green. Figure ***a*** shows a block diagram that pseudocolors an image. The graylevel image F(x, y) is mapped to three images R(x, y), B(x, y) and G(x, y). Each of these images is then used to modulate the red, blue, and green guns of the imaging display's picture tube or CRT.

To display a graylevel image, the mapping functions, R, B, and G, contain the same identical mapping function given by the input graylevel values, F. The three mapping functions can also be used to control the brightness and contrast of the image by modifying all three mapping functions equally.

For example, to reduce the contrast of an image R = B = G = 0.5F. The output intensity values of the three images are one half the graylevel values of the input image.

For example, to highlight a particular graylevel value T in red, the following equation can be used:

$$g(x, y) = \begin{cases} R = F: B = G = 0 & \text{for } F(x, y) = T \\ R = G = B = F & \text{elsewhere} \end{cases}$$

In the above equation, for F(x, y) equal to the desired graylevel value T, both the blue and green images are set to zero or to a low intensity, and the red image is set to the input image's graylevel value. For all other graylevel values, the red, blue, and green images are set equal to the input image's graylevel values, producing a black and white image for these input graylevel values.

ALGORITHM:

The program assumes that the original image is a 256 graylevel × IMAGE->Rows × IMAGE->Cols pixel image stored in the structure IMAGE. The program then performs a 3 × 3 Sobel edge detection of the original image. Upon completion of the program, edges within the image are highlighted in red and stored in the three structures RED GREEN and BLUE. The three color structures are assumed to be of the same size and type as the original image.

```
Pseudo_edge(struct Image *IMAGE, struct
Image *RED, struct Image *BLUE, struct
Image *GREEN, int T)
      {
      int X,Y, x1, y1, mask1[3][3];
      int mask2[3][3];
      int GX, GY, EDGE;
      long int R, R1;
      mask1[0][0]=-1; mask1[1][0]=-2;
      mask1[2][0]=-1;
      mask1[0][1]= 0; mask1[1][1]= 0;
      mask1[2][1]= 0;
      mask1[0][2]= 1; mask1[1][2]= 2;
      mask1[2][2]= 1;
```

```
    mask2[0][0]=-1; mask2[1][0]= 0;
    mask2[2][0]= 1;
    mask2[0][1]=-2; mask2[1][1]= 0;
    mask2[2][1]= 2;
    mask2[0][2]=-1; mask2[1][2]= 0;
    mask2[2][2]= 1;
    for(Y=1; Y<IMAGE->Rows-1; Y++)
       for(X=1; X<IMAGE->Cols-1; X++)
          {
          GX=0; GY=0;
          for(y1=-1; y1<=1; y1++)
             for(x1=-1; x1<=1; x1++)
                {
                R=X+x1+(long)(Y+y1)*
                IMAGE->Cols;
                R1=X+(long)Y*IMAGE->Cols;
                GX += mask1[x1+1][y1+1]*
                *(IMAGE->Data+R);
                GY += mask2[x1+1][y1+1]*
                *(IMAGE->Data+R);
                }
          EDGE=abs(GX)+abs(GY);
          if(EDGE  > T)
            {
            *(RED->Data+R1)=255;
            *(BLUE->Data+R1)=0;
            *(GREEN->Data+R1)=0;
            }
          else
            {
            *(RED->Data+R1)=
            *(IMAGE->Data+R1);
            *(BLUE->Data+R1)=
            *(IMAGE->Data+R1);
            *(GREEN->Data+R1)=
            *(IMAGE->Data+R1);
            }
          }
}
```

SEE ALSO: RGB, HSI and YIQ Color Models, Pseudocolor Display, and CIE Color chart

CLASS: Color Image Processing

DESCRIPTION:

A pseudocolor display is not a true-color display system. It gives the user the capability of choosing a subset of colors from a huge assortment of colors. A pseudocolor display system is shown in Figure ***a***. This imaging system has only one image storage area of size 512 × 512 pixels by 256 graylevels, requiring 262,144 bytes of memory. The output of the image memory is mapped through three Look-Up-Tables (LUTs) of 8 bits each. The LUTs are used to map each of the 256 graylevel values to the colors red, blue, and green. The output of the LUTs drives three digital-to-analog converters that become the red, blue, and green inputs to the color display.

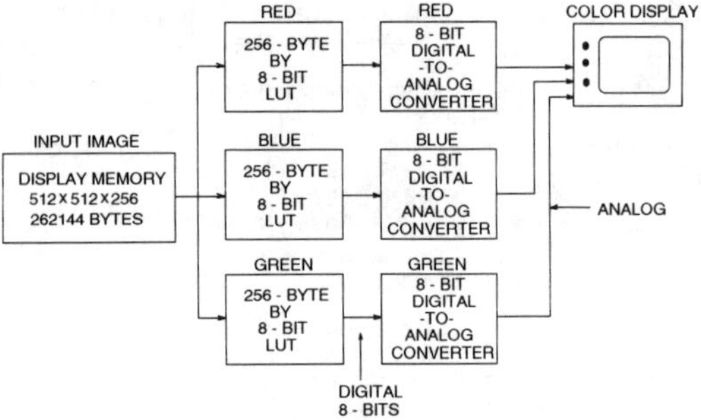

(a) Pseudocolor display system

The LUTs are implemented using three 256 bytes by 8 bits digital read/write memory devices. These memories are used to hold the mapping coefficients for the three primary light colors yielding 16 million possible color combinations.

SEE ALSO: RGB, HSI and YIQ Color Models, C.I.E. Color Chart and True-color Display

CLASS: Image Fundamentals

DESCRIPTION:

Quantization describes the range of values that a *pixel* may take. It is determined by the digitization of the image and how many brightness values can be distiguished by the hardware and ultimately represented by the software. Typically, monochrome images are quantized to 256 graylevels, which is more than adequate for human perception. This means that a pixel value can be represented by a single byte of data. The example shows the result of various quantization levels. The *binary image* is a special case where each pixel takes on one of two values and is represented by black or white. The binary image is used extensively in *morphological filtering* and in *mensuration.*

In the example, the top left picture is 256 levels, top right 16, bottom left is 4, and the bottom right is a *binary*, or 2 level image.

EXAMPLE:

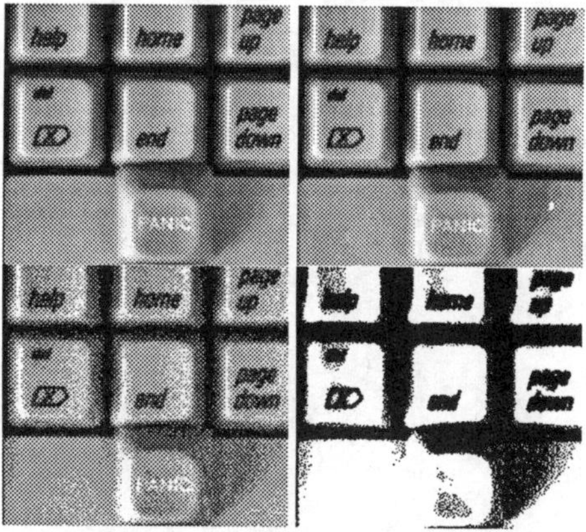

SEE ALSO: Graylevel, Pixel, Sampling, Thresholding

CLASS: Color Image Processing

DESCRIPTION:

The analysis of color has been undertaken by many scientists and engineers for many years. In the seventeenth century, Sir Isaac Newton showed that a beam of sunlight passing through a glass prism emerged as a rainbow of colors. Newton concluded that white light was made from many different colors. Table ***a*** lists the six major color regions and their corresponding wavelengths in nanometers.

Table *(a)* Approximate wavelengths for the six major color regions.

Color	Wavelength
Violet	400 - 450 nm
Blue	450 - 480 nm
Green	480 - 550 nm
yellow	550 - 580 nm
Orange	580 - 610 nm
Red	610 - 700 nm

In the late nineteenth century, Clerk E. Maxwell showed that a color image could be created using three color images. He proposed that three basic colors red (R), blue (B), and green (G), mixed in proportions, were all that were needed to create a color image. Figure ***a*** shows the 3 primary colors of light and their corresponding secondary colors.

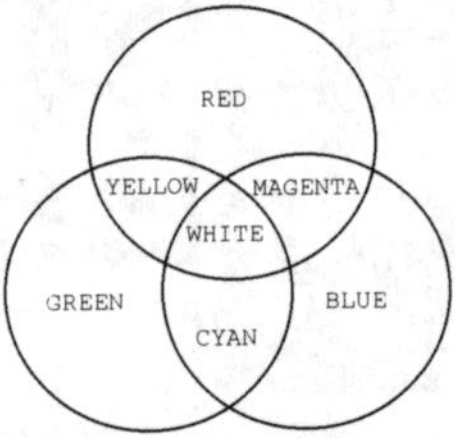

(a) Mixtures of the three primary colors of light.

The percentage of red, blue, and green in a color is known as the color's *trichromatic coefficients*:

$$r = \frac{R}{R+B+G}$$

$$b = \frac{B}{R+B+G}$$

and

$$g = \frac{G}{R+B+G},$$

where R, B, and G are the amount of red, blue, and green light, respectively. The trichromatic coefficients differ from the actual color intensity values R, B, and G in that the trichromatic coefficients have been normalized between 0 and 1. The sum of the three trichromatic coefficients yields

$$r + b + g = 1 .$$

The trichromatic coefficients are computed from the C.I.E. chart (see C.I.E. color chart) from the color's hue and saturation. The x axis gives the red, r, while the y axis gives the green, g, trichromatic coefficients. The blue trichromatic coefficient, b, can then computed using $b = 1 - r - g$. For example, the point labeled red in the C.I.E. chart with a wavelength of 625 nm has the following trichromatic coefficients: r = 71%, b = 0%, and g = 29%, yielding a color combination of 0% white and a saturated color hue of 71% red and 29% green.

SEE ALSO: HSI and YIQ Color Models, C.I.E. Color Chart, and Pseudocolor

CLASS: Nonlinear Filters

DESCRIPTION:

The range filter can be used to find edges within an image. The range filter output is the difference between the maximum and minimum graylevel values within a local region of the image determined by a specified mask. The definition of the range filter is

$$\text{Range}(A) = \max[\, A(x+i, y+j)\,] - \min[\, A(x+i, y+j)\,]\,,$$

where the coordinate x + i, y + j is defined over the image A and the coordinate i, j is defined over the mask M. The mask M determines which pixels are to be included in the range calculation.

EXAMPLE:

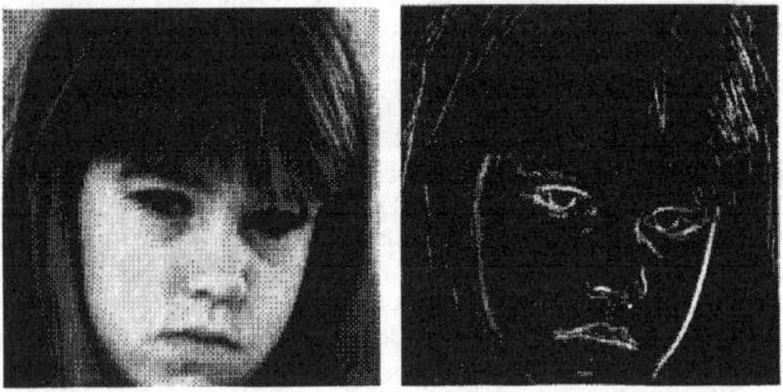

(a) ***(b)***

(a) The original image and ***(b)*** the range filtered image using a 5 × 5 square mask.

ALGORITHM:

The program assumes that the original image is a 256 graylevel × IMAGE->Rows × IMAGE->Cols pixel image stored in the structure IMAGE. The program computes the range filter over a set of pixels contained within a square N × N region of the image centered at the pixel X, Y. The size of the filtering operation is determined by the variable N. The variable N should be set to an odd number and be less than 12. Upon completion of the program, the range filtered image is stored in the structure IMAGE1.

```
Range(struct Image *IMAGE, struct Image
*IMAGE1)
      {
      int X, Y, I, J, smin, smax, N;
      int a[11][11];
      N=3;
      for(Y=N/2; Y<IMAGE->Rows-N/2; Y++)
         for(X=N/2; X<IMAGE->Cols-N/2; X++){
            smin=255; smax=0;
            for(J=-N/2; J<=N/2; J++)
              {
              for(I=-N/2; I<=N/2; I++)
                 {
                  a[I+N/2][J+N/2]=*(IMAGE->
                  Data+X+I+(long)(Y+J)
                 *IMAGE->Cols);
                 }
              }
            for(J=0; J<=N-1; J++)
               {
               for(I=0; I<=N-1; I++)
                  {
                  if(a[I][J] < smin)
                    smin = a[I][J];
                  }
               }
            for(J=0; J<=N-1; J++)
               {
               for(I=0; I<=N-1; I++)
                  {
                  if(a[I][J] > smax)
                    smax = a[I][J];
                  }
               }
            *(IMAGE1->Data+X+(long)Y
            *IMAGE->Cols) = smax - smin;
            }
      }
```

SEE ALSO: Geometric, Y_p, Harmonic, Arithmetic Mean, Median, and other Nonlinear Filters.

CLASS: Noise

DESCRIPTION:

Rayleigh type noise appears typically in radar range and velocity images and is derivable from uniform noise. The Rayleigh noise histogram is defined as

$$h_i = \frac{G_i}{a^2} \exp^{-G_i^2/2a^2} \quad \text{for } 0 \leq G_i < \infty ,$$

where G_i is the ith graylevel value of the image. The mean can be defined in terms of the parameter a as mean $= \sqrt{\pi/2}a$.

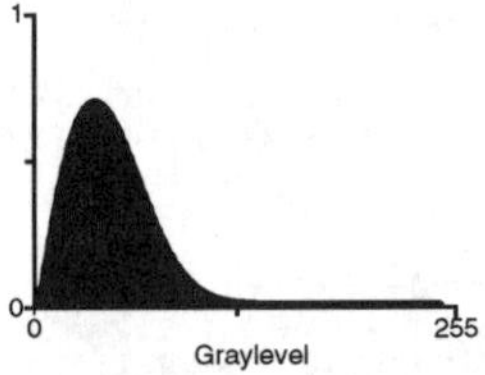

A histogram of Rayleigh noise.

EXAMPLES:

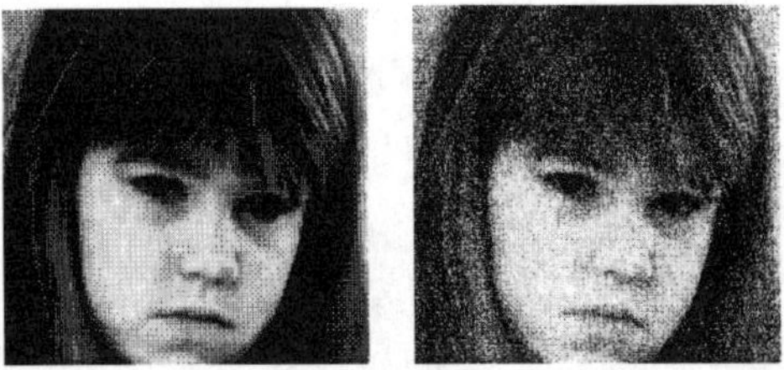

(a) *(b)*

(a) The original image and *(b)* the (additive) Rayleigh noise degraded image with a variance = 600.

ALGORITHM:

The program generates a Rayleigh noise image of 256 graylevel × IMAGE->Rows × IMAGE->Cols pixel image stored in the structure IMAGE. The program assumes the function rand() generates a uniform random number in the range of 0 to 32767. The desired variance is passed to the program upon execution. If the noise graylevel value generated exceeds the 256 graylevel range, the noise graylevel value is truncated to either 0 or 255.

```
Rayleigh(struct Image *IMAGE,
float VAR)
      {
      int X, Y;
      float NOISE, A;
      for(Y=0; Y<IMAGE->Rows; Y++)
         for(X=0; X<IMAGE->Cols; X++)
            {
            A= 2.3299 * VAR;
            NOISE=sqrt(-2 * A * log(1.0-
            (float)rand() / 32767.1));
            if(NOISE > 255)
                  NOISE = 255;
            if(NOISE < 0)
                  NOISE = 0;
            *(IMAGE->Data+X+(long)Y*
            IMAGE->Cols) = (unsigned
            char)(NOISE +.5);
            }
      }
```

SEE ALSO: Gaussian, Uniform, Negative Exponential, Salt and Pepper and Gamma Noises

CLASS: Spatial Filters

DESCRIPTION:

Robert's Filter , or Robert's gradient is a simple 2×2 mask that computes the difference between a pixel and its horizontal and vertical neighbors. The Robert's mask is given below:

0	-1
-1	0

EXAMPLE:

Original Image

Robert's Gradient

ALGORITHM:

The algorithm for Discrete Convolution is applied using the masks given above.

SEE ALSO: Discrete Convolution, High Pass Spatial Filters

CLASS: Graphics Algorithms

DESCRIPTION:

Rotation is used to turn an image. This is often useful when comparing objects in pictures or when generating graphics.

EXAMPLE:

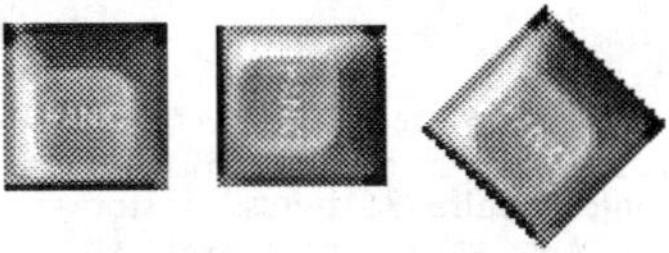

ALGORITHM:

The routine **rotate_90** rotates the image passed by the image data structure pointer **In** by 90° into the image data structure **Out**. More complex rotations, e.g., 45°, such as shown above may be accomplished by the **warp** function. The macro **idx** is used for simplified access to the pixels of the image by coordinates.

```
#define idx(Im,i,j) \
        *(Im->Data + (i)*Im->Cols + (j))

rotate_90(struct Image *In,
          struct Image *Out)
{
   int i,j,k;

   k  = In->Cols;
   for(i = 0; i<In->Rows;++i){
      k  = In->Cols;
      for(j = 0; j<In->Cols;++j)
         idx(Out,k--,i) = idx(In,i,j);
   }
}
```

SEE ALSO: Flip, Warping

CLASS: Coding and Compression

DESCRIPTION:

Run Length Encoding (RLE) is a simple, powerful scheme for the compression of images. In its simplest form, it consists of a repeat count (sometimes called the index) followed by a repeat value. RLE takes advantage of homogeneous areas of an image with long runs of identical pixels.

EXAMPLE:

The 6 × 6 image is of a black background (0 values) with a small white square in the center (255), 8-bit grayscale. This picture would require 36 bytes of storage. The RLE compression (count, value) format, is shown to the right and would require 10 bytes of storage.

0	0	0	0	0	0
0	0	0	0	0	0
0	0	255	255	0	0
0	0	255	255	0	0
0	0	0	0	0	0
0	0	0	0	0	0

→

(14,0)
(2,255)
(4,0)
(2,255)
(14,0)

If the image pixel values are highly uncorrelated, then the RLE encoding can require *more* storage area than the actual data.

ALGORITHM:

The **rle** routine codes or decodes the unsigned character data pointed to by the image structure pointer, **In**. The coded data is written or read from the file, **filename**. The direction flag, **dir**, is set to non-zero for compression, or zero for decoding.

The RLE scheme used is that of paired bytes, where the first byte of the pair identifies the run length and the second byte specifies the run value.

```
/* Run-Length Encoding Algorithm */

rle(struct Image *In,char *filename,
      char dir)
{
   int sz;
   unsigned char *Im,run,val;
   FILE *fp;

   /* Init image size index & data ptr */
   sz = In->Rows * In->Cols;
   Im = In->Data;

   if (dir){  /* COMPRESS */
      fp = fopen(filename,"wb");
      do{
         val = *(Im++);
         --sz;
         run = 1;
         while((val == *Im) && sz){
            ++run;
            ++Im;
            --sz;
            if(run==0xff)break;
         }
         fputc(run,fp);   /* write index */
         fputc(val,fp);   /* write value */
      } while (sz);
   }
   else {    /* DECOMPRESS */
      fp = fopen(filename,"rb");
      do{
         /* get index */
         run = fgetc(fp) & 0xff;
         /* get value */
         val = fgetc(fp) & 0xff;
         *Im++ = val;
         while(--run && --sz)*Im++ = val;
      } while (sz);
   }
   fclose(fp);
}
```

SEE ALSO: PCX, MAC

CLASS: Noise

DESCRIPTION:

Salt and pepper type noise typically occurs in images that are acquired by cameras containing malfunctioning pixels. Salt and pepper noise is named after the white and black appearance it adds to images. Its histogram is defined as

$$h_i = \begin{cases} \text{pepper noise with probability p} & \text{for } G_i = a \\ \text{salt noise with probability p} & \text{for } G_i = b \\ 0 & \text{elsewhere} \end{cases},$$

where G_i is the ith graylevel value of the image. The salt and pepper noises each occur at graylevel values a and b with probability p.

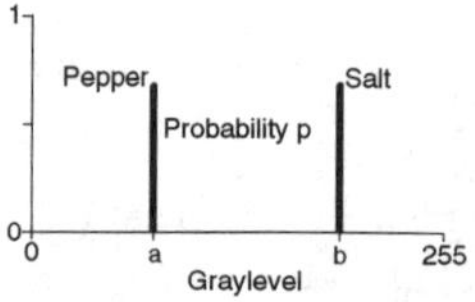

A histogram of salt and pepper noise.

EXAMPLES:

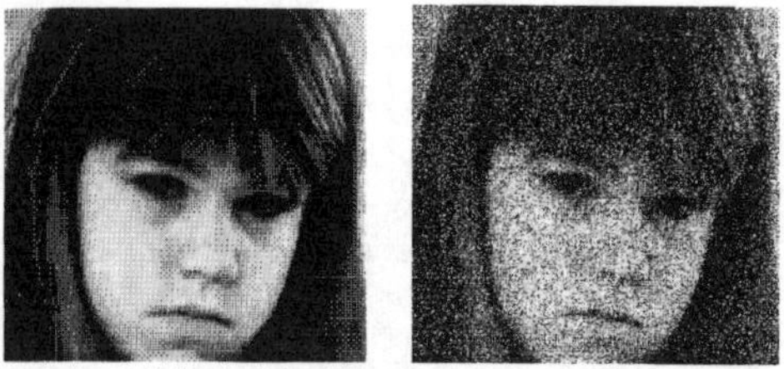

(a) *(b)*

(a) The original image and *(b)* the salt and pepper noise degraded image with a combined probability of 20%.

ALGORITHM:

The program generates a salt and pepper noise image of 256 graylevel × IMAGE->Rows × IMAGE->Cols pixel image stored in the structure IMAGE. The program assumes that both salt and pepper noises are equally likely of occurring. Also, the program assumes the function rand() generates a uniform random number in the range of 0 to 32767. The desired probability of occurrence for both the salt and pepper noises are passed to the program upon execution. The pepper noise is given a graylevel value of 0 while the salt noise is assigned a graylevel value of 255.

```
SaltPepper(struct Image *IMAGE,
float PROBABILITY)
     {
     int X, Y, DATA, DATA1, DATA2;
     float NOISE;
     DATA=(int)(PROBABILITY*32768
     /2);
     DATA1=DATA + 16384;
     DATA2=16384 - DATA;
     for(Y=0; Y<IMAGE->Rows; Y++)
        for(X=0; X<IMAGE->Cols; X++)
          {
          DATA=rand();
          if(DATA>=16384 && DATA<DATA1)
               *(IMAGE->Data+X+(long)Y
               *IMAGE->Cols) = 0;
          if(DATA>=DATA2 && DATA<16384)
               *(IMAGE->Data+X+(long)Y
               *IMAGE->Cols) = 255;
          }
     }
```

SEE ALSO: Gaussian, Uniform, Negative Exponential, Rayleigh and Gamma Noises

CLASS: Image Fundamentals

DESCRIPTION:

Sampling describes the spatial distribution of the image acquisition. We assume that pixels ajoin each other with no space between them. The ratio of the number of pixels in a row or column with the width or height of the image defines the spatial frequency, or sample rate. Images are generally sampled at binary multiples, such as 256 × 256, 512 × 512, or 1024 × 1024. The resolution of an image, or how much spatial detail may be resolved, is largely determined from the sample rate. In the example below, the same image is shown with two different sample rates, and different sized pixels. If the pixels remain the same size, the image with more pixels will show greater detail. Images are typically represented as matrices in the computer, although more elaborate data structures are possible.

EXAMPLE:

Image sampled at 64 × 64 pixels. Image at 256 × 256 pixels.

SEE ALSO: Pixel, Quantization, Spatial Frequency

CLASS: Image Fundamentals

DESCRIPTION:

Scaling as discussed here refers to min-max scaling of the result of an image process to a quantization scale. For example, the result of an algorithm will often be an image with floating point values. To output this image, the values must often be scaled to a particular *grayscale*, most often within the range of 0-255. The algorithm given accepts a floating point input image and outputs an image where each pixel is represented by a scaled byte (character) value. The formula for scaling a 256 graylevel image is:

$$\text{OutPixel} = \frac{255}{\text{max-min}}(\text{InPixel - min})$$

where the max and min values are the global maximum and minimum value of all pixels in the input image.

ALGORITHM:

The algorithm accepts two image structures of type float and no checking is made to verify this. The initializing values of the *min* and *max* test variables, FLT_MAX and FLT_MIN, should be predefined by the system being used, but for this example have been set to arbitrary values. Look for these, or similar defines, in the math.h, limits.h, and float.h include files on your system. The **InD** and **OuD** float pointers are used as convenience indexes for the scale evaluation loops.

If the data is of exceedingly large range, then an exponential scaling can be applied by taking the log of the input image values prior to scaling.

```
#define FLT_MAX  99999999
#define FLT_MIN -99999999

/* Scale image */
void Scale(struct Image *In,
           struct Image *Out)
{
   float min, max, *InD, *OuD;
   long ImSize, i;

   min = FLT_MAX;
   max = FLT_MIN;
```

```
    InD = (float *)In->Data;
    ImSize = In->Rows * In->Cols;

    /* scan input for min/max values */
    for(i=0;i<ImSize;++i){
          if(*InD < min)
             min = *InD;
          if(*InD > max)
             max = *InD;
          ++InD;
    }

    InD = (float *)In->Data;
    OuD = (float *)Out->Data;

    /* scale the Output */
    for(i=0;i<ImSize;++i)
          *OuD=(255/(max-min))*(*InD-min);
}
```

SEE ALSO: Discrete Convolution, Discrete Correlation, Graylevel, Histogram Operations

Segmentation

DESCRIPTION:

An important area of image processing is the segmentation of an image into various components for object recognition. The goal of segmentation is to separate the object's pixels within an image from the background pixels. Thresholding techniques separate an object from the background based upon the graylevel histogram of an image. If the graylevel values of an object within an image are quite different than the background graylevel value, then finding the optimum threshold value to threshold the image is quite simple.

Other segmentation techniques use the graylevel discontinuities within an image to detect lines or points within the image. These discontinuities are then used to separate objects within an image from the background.

CLASS MEMBERSHIP:

- Line Detector
- Multi-Graylevel Thresholding
- Optimum Thresholding
- Point Detector
- Thresholding

CLASS: Morphological Filters

DESCRIPTION:

Skeletonization of an object, initially proposed by H. Blum, defines a unique compressed geometrical representation of an object. Skeletonization of an object is often referred to as the Medial Axis Transform. Morphological skeletonization is defined as the union of the set of pixels computed from the difference of the nth eroded image and the opening of the nth eroded image.

$$K_n(A) = \text{Erode}_n(A) - \text{open}(\text{Erode}_n(A), B)\ ,$$

where $\text{Erode}_n(A) = A \ominus nB$ and is the nth erosion of the original image A with the structuring function B.

The skeleton image is then given by the union of all $K_n(A)$ over all erosions. The total number of erosions N required by the skeleton algorithm is the number of erosions of the original image A by the structuring function **B** that yields the null image.

$$\varnothing = \text{Erode}_N(A) = A \ominus NB$$

EXAMPLES:

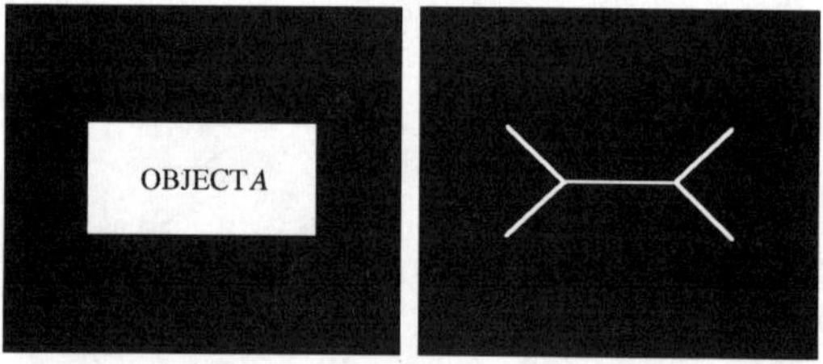

(a) *(b)*

(a) The original binary image and ***(b)*** the skeleton image.

(a)

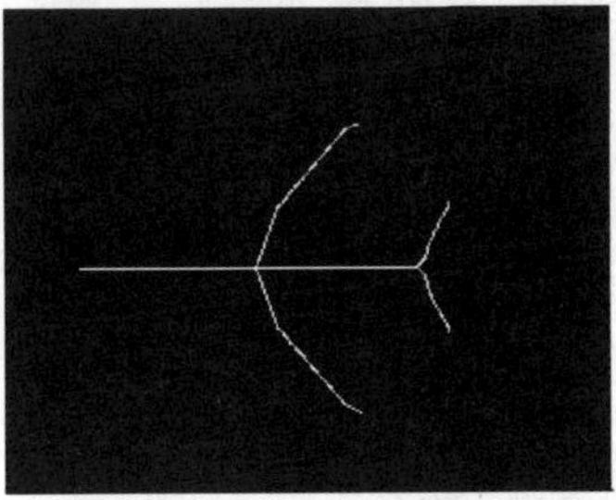

(b)

(a) The original binary image of an airplane silhouette and ***(b)*** the skeleton image.

ALGORITHM:

The program assumes that the original binary image is an IMAGE->Rows × IMAGE->Cols pixel image with a background graylevel value of 0 and an object graylevel value of 255 (object) and is stored in the structure IMAGE. A N × N structuring function is used by the algorithm and is stored in array MASK[][]. Upon completion of the program, the skeleton image is stored in the structure SKELETON. The erosion and dilation functions used by the algorithm can be found under binary erosion and dilation respectively.

```
#define N 3

Skeleton(struct Image *IMAGE,int MASK[][N],
        struct Image *SKELETON)
      {
      int X, Y, I, J;
      int pixel_on, false, true, pixel;
      struct Image *FILTER, *FILTER1, A, A1;
      FILTER=&A;
      FILTER1=&A1;
      /* Use these 2 lines for Non MS-DOS
      systems*/
      FILTER->Data=(unsigned char *)
      malloc((long)IMAGE->Cols*(long)
      IMAGE->Rows);
      FILTER->Data=(unsigned char *)
      malloc((long)IMAGE->Cols*(long)
      IMAGE->Rows);
      /*Use these 2 lines for MS-DOS systems
      */
      /*FILTER->Data=(unsigned char huge *)
      farmalloc((long)IMAGE->Cols*(long)
      IMAGE->Rows);
      FILTER1->Data=(unsigned char huge *)
      farmalloc((long)IMAGE->Cols*(long)
      IMAGE->Rows);*/
      FILTER->Rows=IMAGE->Rows;
      FILTER->Cols=IMAGE->Cols;
      FILTER1->Rows=IMAGE->Rows;
      FILTER1->Cols=IMAGE->Cols;
      true=1; false=0; pixel_on = true;
      for(Y=N/2; Y<IMAGE->Rows-N/2; Y++)
         for(X=N/2; X<IMAGE->Cols-N/2; X ++)
            *(SKELETON->Data + X
            +(long)Y*IMAGE->Cols)=0;
      while(pixel_on == true)
          {
          pixel_on=false;
          Erosion(IMAGE, MASK, FILTER);
          Dilation(FILTER, MASK,FILTER1);
          for(Y=N/2; Y<IMAGE->Rows-N/2; Y++)
```

```
        for(X=N/2; X<IMAGE->Cols-N/2; X++)
                {
                pixel=*(IMAGE->Data + X
                +(long)Y*IMAGE->Cols)-
                *(FILTER1->Data + X
                +(long)Y*IMAGE->Cols);
                *(SKELETON->Data + X
                +(long)Y*IMAGE->Cols)=
                *(SKELETON->Data + X
                +(long)Y*IMAGE->Cols) |
                pixel;
                if(pixel==255)
                  pixel_on=true;
                *(IMAGE->Data + X
                +(long)Y*IMAGE->Cols)=
                  *(FILTER->Data + X
                  +(long)Y*IMAGE->Cols);
                }
        }
    }
```

SEE ALSO: Binary Erosion, Dilation, Opening, and Closing Filters

CLASS: Transforms

DESCRIPTION:

The *Slant Transform* uses sawtooth waveforms as a basis set. The transform is symmetric and real and so is easily implemented using matrix cross product multiplication. The transform matrix may be defined recursively from the following expressions:

$$S_2 = \frac{1}{\sqrt{2}}\begin{bmatrix} 1 & 1 \\ 1 & -1 \end{bmatrix}$$

$$S_N = \begin{bmatrix} \begin{bmatrix} 1 & 0 \\ a_N & b_N \end{bmatrix} & \underline{0} & \begin{bmatrix} 1 & 0 \\ -a_N & b_N \end{bmatrix} & \underline{0} \\ \underline{0} & I_{(N/2)-2} & \underline{0} & I_{(N/2)-2} \\ \begin{bmatrix} 1 & 0 \\ -b_N & a_N \end{bmatrix} & \underline{0} & \begin{bmatrix} 0 & -1 \\ b_N & a_N \end{bmatrix} & \underline{0} \\ \underline{0} & I_{(N/2)-2} & \underline{0} & I_{(N/2)-2} \end{bmatrix} \cdot \begin{bmatrix} S_{N/2} & \underline{0} \\ \underline{0} & S_{N/2} \end{bmatrix}$$

$$\mathbf{a}_{2N} = \left(\frac{3N^2}{4N^2-1} \right)^{\frac{1}{2}} \quad \mathbf{b}_{2N} = \left(\frac{N^2-1}{4N^2-1} \right)^{\frac{1}{2}},$$

where $N=2^n$

and the tranform itself is then just the original image multiplied by the transform matrix and the transpose of the transform matrix as follows:

$$F_{slant} = S_{slant}\ [f]\ S'_{slant}$$

with the inverse:

$$[f] = S'_{slant}\ F_{slant}\ S_{slant}$$

The simplest realization of the transform is to have precomputed slant matrices as there is no direct method of computing them, then use matrix utilities to process the transform. The advantage of the slant transform over others is the closer approximation one derives to the optimal basis set expansion without entering the complex domain.

The transform represents coefficients of sequency, as with the Walsh-Hadamard transforms.

ALGORITHM:

The **Slant** algorithm assumes that a slant matrix has been computed for the size N × N input image, **In**. This matrix may be computed using the routine **slant_matrix**, given below. This routine computes the data values for the **Slant** image structure and should be run offline and the slant matrix stored. If the direction variable **d** passed to the transform routine is nonzero, the forward transform is computed, if zero, the inverse transform is processed. The image utilities, **Multiply**, **Transpose** and **Copy** form the core of the transform and may be used with other multiplicative kernels such as the Walsh and Hadamard matrices. In fact, the image structure **Slant** may be filled with either of those kernels and the routine will yield those transforms.

The user should be aware that these routines allocate memory as needed and no error checking is performed. This is particularly important in the **slant_matrix** routine as it is recursive and generates a large number of matrices. Also, the routines use access macros, specifically the **Multiply** routine. for easy cartesian access to the matrix arrays.

```
/* Slant transform */
/* calls Multiply, Transpose and Copy
   routines */
Slant(struct Image *In, struct Image *Slant,
      struct Image *Out, char d)
{
   struct Image Scratch;
   float tmpf;

   Scratch.Rows = Scratch.Cols = In->Rows;
   Scratch.Data = (unsigned char *)
      calloc(In->Rows,sizeof(tmpf)*In->Cols);
```

```
   if(d){
      Multiply(Slant,In,Out);
      Transpose(Slant,&Scratch);
      Copy(&Scratch,Slant);
      Copy(Out,&Scratch);
      Multiply(&Scratch,Slant,Out);
   }
   else{
      Multiply(In,Slant,Out);
      Transpose(Slant,&Scratch);
      Copy(&Scratch,Slant);
      Copy(Out,&Scratch);
      Multiply(Slant,&Scratch,Out);
   }
}

/* compute the recursive a(N) coefficient
*/
float SLa(int N){
   extern float SLb(int N);

   if(N==2)return(1.0);
   return(2.0*SLb(N)*SLa(N/2));
}

/* compute the recursive b(N) coefficient
*/
float SLb(int N){
   return(1.0/
         sqrt(1.0+4*(SLa(N/2)*SLa(N/2))));
}


#define idx(i,j)    *(tmp + (i)*Scr.Cols + j)

void slant_matrix(struct Image *SLM, int N)
{
   float K2, *tmp,*tmpR;
   struct Image Scr,Scr2,SRM;
   long NI,i,j;

   K2 = 1.0/sqrt(2.0);

   if(N==2){
      SLM->Rows = 2;
      SLM->Cols = 2;
      SLM->Data = (unsigned char *)
                  calloc(4,sizeof(K2));
```

```
      tmp = (float *)SLM->Data;
      *tmp++ = K2;
      *tmp++ = K2;
      *tmp++ = K2;
      *tmp   = -1.0*K2;

      return;
   }

   Scr.Rows = N;
   Scr.Cols = N;
   Scr.Data = (unsigned char *)
              calloc(N*N,sizeof(K2));
   NI = N/2 - 2;

   tmp = (float *)Scr.Data;

   idx(0,0) = K2;
   idx(0,1) = 0.0;
   idx(1,0) = SLa(N);
   idx(1,1) = SLb(N);

   idx(0,2+NI) = K2;
   idx(0,3+NI) = 0.0;
   idx(1,2+NI) = -1.0*SLa(N);
   idx(1,3+NI) = SLb(N);

   idx(2+NI,0) = 0.0;
   idx(3+NI,0) = -1.0*SLb(N);
   idx(2+NI,1) = K2;
   idx(3+NI,1) = SLa(N);

   idx(2+NI,2+NI)  = 0.0;
   idx(2+NI,3+NI)  = -1.0*K2;
   idx(3+NI,2+NI)  = SLb(N);
   idx(3+NI,3+NI)  = SLa(N);

   if(NI){
      j=2*NI;
      for(i=2; i<2+NI; ++i){
         idx(i,i)     = K2;
         idx(i,i+j)   = K2;
         idx(i+j,i)   = K2;
         idx(i+j,i+j) = K2;
      }
   }
```

```
   /* Compute recursive multiplier */
   slant_matrix(&Scr2,N/2);

   /* Generate the Recursive Multiplier */
   SRM.Rows = N;
   SRM.Cols = N;
   SRM.Data = (unsigned char *)
              calloc(N*N,sizeof(K2));

   tmp  = (float *)SRM.Data;
   tmpR = (float *)Scr2.Data;
   for(i=0;i<N/2;++i){
      for(j=0;j<N/2;++j)
         *tmp++ = *(tmpR + i*Scr2.Rows + j);
      tmp += N/2;
   }
   tmp  = (float *)SRM.Data;
   tmp  += (N/2)*SRM.Rows + N/2;
   tmpR = (float *)Scr2.Data;
   for(i=0;i<N/2;++i){
      for(j=0;j<N/2;++j)
         *tmp++ = *(tmpR + i*Scr2.Rows + j);
      tmp += N/2;
   }
   SLM->Rows = N;
   SLM->Cols = N;
   SLM->Data = (unsigned char *)
               calloc(N*N,sizeof(K2));
   Multiply(&Scr,&SRM,SLM);
}

#define Z(i,j)  *(OD + (i)*sz + j)
#define A(i,j)  *(AD + (i)*sz + j)
#define B(i,j)  *(BD + (i)*sz + j)

/* Image Multiply (Cross-Product) */
void Multiply(struct Image *A,
              struct Image *B,
              struct Image *Out)
{
   long i,j,k,sz;
   float *AD, *BD, *OD;

   sz = A->Rows;

   AD = (float *)A->Data;
   BD = (float *)B->Data;
   OD = (float *)Out->Data;
```

```
   for(i=0;i<sz*sz;++i)*(OD++)=0.0;
   OD = (float *)Out->Data;

   for(i=0; i < sz; ++i)
      for(j=0; j < sz; ++j)
         for(k=0; k < sz; ++k)
            Z(i,j) = Z(i,j) + A(i,k)*B(k,j);
}

/* Image Transpose */
void Transpose(struct Image *In,
               struct Image *Out)
{
   long i,j;
   float *OD,*ID;

   OD = (float *)Out->Data;
   ID = (float *)In->Data;

   for(i=0; i< In->Rows; ++i)
      for(j=0; j< In->Cols; ++j)
         *(OD++) = *(ID + i + j*In->Cols);

}

/* Image Copy */
void Copy(struct Image *In,
          struct Image *Out)
{
   long i;
   float *OD,*ID;

   OD = (float *)Out->Data;
   ID = (float *)In->Data;

   for(i=0; i< In->Rows*In->Cols; ++i)
         *(OD++) = *(ID++);

}
```

SEE ALSO: Walsh Transform, Hadamard Transform

CLASS: Spatial Filters

DESCRIPTION:

The *Sobel Filter* is an edge detector whose results yield the magnitude and direction of edges by applying the horizontal and vertical line enhancement masks given below:

-1	-2	-1
0	0	0
1	2	1

Horizontal (G_H)

-1	0	1
-2	0	2
-1	2	1

Vertical (G_V)

The formulas for edge magnitude and phase are given by

$$M_{sobel} = \sqrt{G_H^2 + G_V^2} \qquad \phi_{sobel} = \tan^{-1}\left(\frac{G_V}{G_H}\right)$$

A simpler form for M_{sobel} using absolute values is given by

$$M_{sobel} = |G_H| + |G_V|$$

and may be used in place of the radical form.

The image formed from the Sobel magnitude expression yields the edge detection. An image formed from the phase computation is visually interesting, but difficult to interpret. Each pixel value represents not intensity, but edge direction (units are dependent on scale factors and how the inverse tangent was computed).

EXAMPLE:

Original

Sobel Magnitude (absolute value form)

ALGORITHM:

The algorithm for Discrete Convolution is applied using the horizontal and vertical masks given above and the scaled results processed for magnitude and direction.

SEE ALSO: Discrete Convolution, High Pass Spatial Filters

Spatial Filters Class

DESCRIPTION:

The Spatial Filters are basically *discrete convolution* filters or filters that convolve one image with another. The filter image is typically very small with respect to the target image and is called a *spatial mask*. The simplest definition for convolution without resorting to complex mathematical constructs is that it is an operation that copies one image at each pixel location of another while allowing for the effects of all pixel values in the area where the copy takes place. This is accomplished by a multiplying, adding, and shifting operation, hence the term *convolve*, which means to roll, twist, or coil together. Convolution can occur when signals, such as images, are modified by optical, electronic, or nervous systems. The modification of signals in this way yields desirable results when the outcome extracts information that would not be obtainable otherwise, such as in an edge-detection process. Undesirable results occur when a signal is distorted by a defocused lens, and this too is a convolution process.

Many different spatial filter masks can be generated for a variety of functions. The spatial filtering process is especially attractive to image processing because of the computational simplicity of the discrete convolution algorithm and the easy extension of the process into parallel architectures for extremely fast execution.

CLASS MEMBERSHIP:

- Gaussian Filters
- Gradient Masks
- High Pass Spatial Filters
- Laplacian
- Low Pass Spatial Filters
- Robert's Filter
- Sobel Filter
- Spatial Masks

CLASS: Image Fundamentals

DESCRIPTION:

Spatial Frequency is a measure of the periodicity of a two dimensional data set with respect to a distance measure. Periodic changes in brightness values across an image are defined in terms of spatial frequency, or periods/distance. If the period of a brightness pattern is 300 pixels and the size of a pixel is 1/150th of an inch, then the spatial frequency of the data set is 2 cycles/inch.

The black and white lines in the example of low spatial frequency below are 25 pixels wide, giving a period of 50 pixels. A complete period starts with a black line and ends at the next black line, moving horizontally. The lines are printed at 150 pixels/inch, thus the spatial frequency of the lines is 2 cycles/inch. The high spatial frequency lines to the right are approximately half as wide with a 4 cycle/inch frequency.

The *Discrete Fourier Transform* resolves the spatial frequency components of an image. In complex images, low spatial frequencies are characterized by slow, broad changes in brightness, like the wide lines in the example below. High spatial frequencies occur when there are abrupt, sharp changes in brightness, such as when an image contains fine lines and sharp edges, also illustrated in the example.

EXAMPLE:

Low Spatial Frequency

High Spatial Frequency

SEE ALSO: Discrete Fourier Transform, Quantization, Sampling

Spatial Frequency Filters Class

DESCRIPTION:

Image processing filters within this class operate directly on the spatial frequency decomposition of the image, unlike spatial filters, in the spatial frequency domain. To do this, the image must first be transformed into a frequency domain representation, and this is most often accomplished through use of the *Fourier Transform.*

Filters within this class are most often used for restoration purposes (the exception being the circularly symmetric and homomorphic filters, which are used for enhancement). Restoration algorithms seek to remove a degradation and/or noise that has corrupted the image. The general model assumed for restoration purposes is given by the following block diagram:

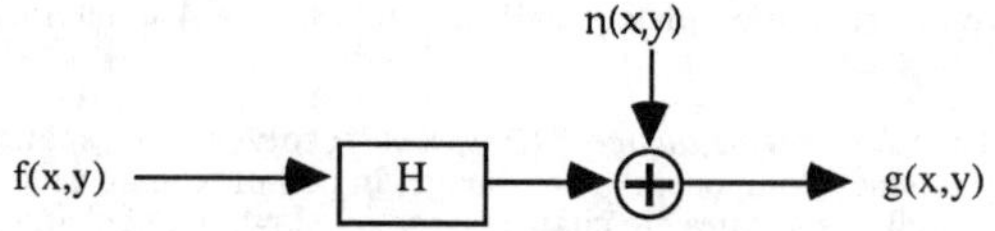

Here an image, f(x,y), is convolved with a degradation function, H, followed by the addition of a noise function, n(x,y), to yield a noisy, degraded picture, g(x,y).

CLASS MEMBERSHIP:

Circularly Symmetric Filter

Homomorphic Filter

Inverse Filter

Least Mean Squares Filter

Parametric Wiener Filter

Wiener Filter

SEE ALSO: Spatial Filters, Fourier Transform

Spatial Masks

CLASS: Spatial Filters

DESCRIPTION:

Spatial Masks are arrays of numbers that are applied to images using the Discrete Convolution Algorithm. They are the basis of spatial filters and are sometimes called windows or frames. The application of a mask is a computionally intensive operation, hence masks are generally 3×3 arrays. Application of a mask is a neighborhood process and issue exists as to which element of the neighborhood will serve as the replacement point of the computation. Typically, masks contain a center value and have an odd number of elements on each side. This is by no means a requirement (see *Robert's Filter*), but adhering to the convention makes computation simpler. Hence, one often sees 3×3, 5×5, and 9×9 masks, as shown in the example below.

EXAMPLE:

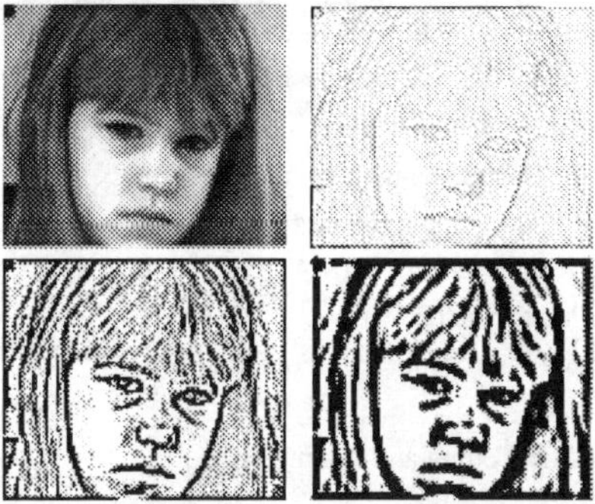

Original　　　3×3 Laplacian

5×5 Laplacian　　　9×9 Laplacian

Storage Formats Class

DESCRIPTION:

Images contain large amounts of data that is derived from a wide range of sources in various formats. It is only natural that specialized data formats were created to handle the storage, transmission, and exchange of image data. The number of image storage formats available is nearly as numerous as the number of image acquisition and display hardware devices available. In the early days of computer image processing, each imaging system had a unique way of representing the image data that it created and used. As networking became popular, it became important (1) that images could move easily from platform to platform, and (2) that image data be compressed so as to minimize the time and storage required for these huge data entities. Format schemes became popular either because a large number of desirable images were formatted using it or because it was used by a popular hardware or software system. In any event, this class includes five well-known image file format techniques.

CLASS MEMBERSHIP:

- Graphics Interchange Format (GIF)
- Joint Photographic Experts Group (JPEG)
- MacPaint File Format (MAC)
- PC Paintbrush (PCX)
- Tagged Interchange File Format (TIF)

SEE ALSO: Coding and Compression

CLASS: Storage Formats

DESCRIPTION:

The *Tagged Interchange File Format*, or TIF, is an image storage format based on a sequence of individual fields defined by unique *tags*. These tags are descriptors, or pointers, to defined fields containing data and data descriptions of an image. This pointer structure allows TIF to describe virtually any type of image data and be extensible when new formats are developed. A TIF file consists of a sequence of up to 2^{32} bytes that a reader program interprets as a tag that tells it what to do with the data that follows, or as specific data associated with a description identified by the tag, including compressed or encoded data.

The TIF file consists of an 8-byte header followed by one or more image file directories. The data contained in the directories are identified by their tags which are read as 2-byte integers. This allows for up to 65,535 different tags. Fortunately, there are not that many defined--yet!

The first two bytes of the file define whether Intel *little-endian* or Motorola *big-endian* byte word ordering is to be used when accessing the files data. If these bytes are 'II' (0x4949), then Intel is specified and the order of a word is from least significant to most significant. If the first byte is 'MM' (0x4D4D), then word ordering is from most significant to least. This holds for both 2 and 4 byte words. Character strings are stored sequentially.

The next two bytes of the file are the version number. This is most often 42 (0x002A or 0x2A00) and may be ignored.

The last four bytes of the header indicate the offset into the file, in bytes, of the first image directory. This is most often 8, indicating that the first directory follows the header.

The Image File Directory (IFD) is a variable length data structure that consists of a 2-byte count followed by *count* 12-byte IFD fields, followed by the 4-byte offset to the next IFD, or zero if the IFD is the last one in the file.

The first two bytes of the IFD field specify the Tag for the field. The next two bytes are the field Type (length in bytes is given in parenthesis): 1=BYTE(1), 2=ASCII(1),

3=SHORT(2), 4=LONG(4), and 5=RATIONAL(8). The RATIONAL type consists of two LONG values, the first being the integer numerator of a fraction and the second the denominator. Following the Type is a 2-byte integer that specifies the Count or how many values of length Type are in the IFD. Finally, the last four bytes contain the Value Offset of the IFD data. The data may be anywhere in the file.

The value of the Tag indicates how the data in the IFD is to be interpreted. A selection of TIF tag values is included in Appendix C for reference purposes.

ALGORITHM:

It is well beyond the scope of this book to provide a complete TIF evaluation program; however, we show a simple TIF file evaluator that indicates whether a file is TIF, then lists the important format data and Tags that are present in the file. This can provide the basis for a reader that then evaluates each of the tags as they appear and according to their specification. The routine **TIF_eval** accepts a pointer, **fp**, to an opened file and evaluates the file for TIF data. If the file is not TIF, a negative value is returned. The **rword** and **rlong** routines are called to evaluate the int and long int tag data according to the data storage convention used in the file. These conventions are determined from the first two bytes of a TIF file, either II for Intel and *little endian*, or MM for Motorola and *big endian*. The routines reorder the data read based on the system used to store it. The actual platform that the routine is run on is unimportant.

```
/* TIF tag types */
char *Typer[6] ={"",
  "BYTE","ASCII","SHORT","LONG","RATIONAL"};

TIF_eval(FILE *fp)
{
   unsigned char buf[128];
   unsigned int rword(FILE *fp,char kind);
   unsigned long rlong(FILE *fp,char kind);
   unsigned int Version,Tag, Type, Ent;
   unsigned long offset, Count, Voffset;
   char kind;
   int i;
```

```
   fread(buf,2,1,fp); /* byte order */

   /* here we set the byte order flag for
      integer reads */

   switch(buf[0]){

      case 'I':     /* intel */
         kind = 1;
         printf("Byte-order is Intel.\n");
         break;

      case 'M':     /* motorola */
         kind = 0;
         printf("Byte-order is Motorola.\n");
         break;

      default:
         printf("Not TIF file!\n");
         return(-1);
   }
   Version = rword(fp,kind);
   printf("Version:%d\n",Version);

   /* get offset & move to IFD start */
   offset = rlong(fp,kind);
   fseek(fp,offset,0);

   Entries = rword(fp,kind);
   printf("%d entries found in IFD\n", Ent);

   /* loop through the fields */
   for(i=0;i<Ent;++i){

      Tag     = rword(fp,kind);
      Type    = rword(fp,kind);
      Count   = rlong(fp,kind);
      Voffset = rlong(fp,kind);

      printf("\nTAG:\t%d\n",Tag);
      printf("Typ:\t%s\n",Typer[Type]);
      printf("Count:\t%lu\n",Count);
      printf("Offset:\t%lx\n",Voffset);
   }

   fclose(fp);
}
```

```
unsigned long rlong(FILE *fp,char protocol)
{
   /* protocol 1 is Intel else Motorola */
   if(protocol)
     return((unsigned long)
      (fgetc(fp)&0xff) +
           ((unsigned long)
      (fgetc(fp)&0xff)<<8)+
           ((unsigned long)
      (fgetc(fp)&0xff)<<16)+
           ((unsigned long)
      (fgetc(fp)&0xff)<<24));
   else
     return(((unsigned long)
      (fgetc(fp)&0xff)<<24)+
           ((unsigned long)
      (fgetc(fp)&0xff)<<16)+
           ((unsigned long)
      (fgetc(fp)&0xff)<<8)+
            (unsigned long)
      (fgetc(fp)&0xff));
}

unsigned int rword(FILE *fp,char protocol)
{
   /* protocol 1 is Intel else Motorola */
   if(protocol)
      return((fgetc(fp)&0xff)+
            ((fgetc(fp)&0xff)<<8));
   else

      return(((fgetc(fp)&0xff)<<8)+
      (fgetc(fp)&0xff));
}
```

SEE ALSO: PC Paintbrush (PCX)

CLASS: Morphological Filters

DESCRIPTION:

The binary thickening operation is the dual of binary thinning and is used to increase the geometric size of an object. Usually this operation is used iteratively until the desired image is obtained. The thickening operation is defined as

for i equal 1 to 8

$$A^{i+1} = A^i \cup [\text{HitMiss}(A^i, B^i, C^i)]$$

Thickening(A) $=A^9$

where A^1 is the original image, A^9 is the thickened image, and B^i and C^i are the ith set of eight masks given by

B^1

1	1	1
0	0	0
0	0	0

C^1

0	0	0
1	1	1
1	1	1

B^2

0	0	0
0	0	0
1	1	1

C^2

1	1	1
1	1	1
0	0	0

B^3

1	0	0
1	0	0
1	0	0

C^3

0	1	1
0	1	1
0	1	1

B^4

0	0	1
0	0	1
0	0	1

C^4

1	1	0
1	1	0
1	1	0

B^5

0	1	1
0	0	1
0	0	0

C^5

1	0	0
1	1	0
1	1	1

B^6

0	0	0
1	0	0
1	1	0

C^6

1	1	1
0	1	1
0	0	1

B^7

0	0	0
0	0	1
0	1	1

C^7

1	1	1
1	1	0
1	0	0

B^8

1	1	0
1	0	0
0	0	0

C^8

0	0	1
0	1	1
1	1	1

EXAMPLES:

(a) ***(b)***

(a) The original image of 3 squares and ***(b)*** the thickened image.

ALGORITHM:

The program assumes that the original binary image is an IMAGE->Rows × IMAGE->Cols pixel image with a background graylevel value of 0 and an object graylevel value of 255 (object) and is stored in the structure IMAGE. Upon completion of the program the thickened image is stored in the THICKEN structure. The number of iterations performed by the program is given by the variable MAXIT. The binary hit-miss function used by the algorithm can be found under binary hit-miss.

```
#define N 3

Thickened(struct Image *IMAGE,
      struct Image *THICKEN, int MAXIT)
      {
      int X,Y,I,J,Z, M1[3][3][8], M4[3][3];
      int M2[3][3][8], stpflg=0, M3[3][3];
      long int R;
      struct Image *FILTER, *IMAGEC, A, A1;
      FILTER=&A;
      IMAGEC=&A1;
      /* Use these 2 lines for Non MS-DOS
      systems*/
      IMAGEC->Data=(unsigned char *)
      malloc((long)IMAGE->Cols*(long)
      IMAGE->Rows);
      FILTER->Data=(unsigned char *)
      malloc((long)IMAGE->Cols*(long)
      IMAGE->Rows);
      /*Use these 2 lines for MS-DOS systems
      */
      /*IMAGEC->Data=(unsigned char huge *)
      farmalloc((long)IMAGE->Cols*(long)
      IMAGE->Rows);
      FILTER->Data=(unsigned char huge *)
      farmalloc((long)IMAGE->Cols*(long)
      IMAGE->Rows);*/
      IMAGEC->Rows=IMAGE->Rows;
      IMAGEC->Cols=IMAGE->Cols;
      FILTER->Rows=IMAGE->Rows;
      FILTER->Cols=IMAGE->Cols;
      R=IMAGE->Cols;
      for(Z=0; Z<=7; Z++)
         for(J=0; J<=2; J++)
            for(I=0; I<=2; I++)
```

```
            M1[I][J][Z]=0;
      M1[0][1][0]=1; M1[1][1][0]=1;
      M1[2][1][0]=1;
      M1[0][2][0]=1; M1[1][2][0]=1;
      M1[2][2][0]=1;
      M1[0][0][1]=1; M1[1][0][1]=1;
      M1[2][0][1]=1;
      M1[0][1][1]=1; M1[1][1][1]=1;
      M1[2][1][1]=1;
      M1[0][0][2]=1; M1[1][0][2]=1;
      M1[0][1][2]=1;
      M1[1][1][2]=1; M1[0][2][2]=1;
      M1[1][2][2]=1;
      M1[1][0][3]=1; M1[2][0][3]=1;
      M1[1][1][3]=1;
      M1[2][1][3]=1; M1[1][2][3]=1;
      M1[2][2][3]=1;
      M1[0][0][4]=1; M1[0][1][4]=1;
      M1[1][1][4]=1;
      M1[0][2][4]=1; M1[1][2][4]=1;
      M1[2][2][4]=1;
      M1[0][0][5]=1; M1[1][0][5]=1;
      M1[2][0][5]=1;
      M1[0][1][5]=1; M1[1][1][5]=1;
      M1[0][2][5]=1;
      M1[0][0][6]=1; M1[1][0][6]=1;
      M1[2][0][6]=1;
      M1[1][1][6]=1; M1[2][1][6]=1;
      M1[2][2][6]=1;
      M1[2][0][7]=1; M1[1][1][7]=1;
      M1[2][1][7]=1;
      M1[0][2][7]=1; M1[1][2][7]=1;
      M1[2][2][7]=1;
      for(Z=0; Z<=7; Z++)
        for(J=0; J<=2; J++)
          for(I=0; I<=2; I++)
          M2[I][J][Z]=1-M1[I][J][Z];
      while(stpflg<MAXIT)
       {
       for(Z=0; Z<=7; Z++)
         {
         for(J=0; J<=2; J++)
           for(I=0; I<=2; I++)
             {
             M3[I][J]=M2[I][J][Z];
             M4[I][J]=M1[I][J][Z];
             }
```

```
      for(Y=0; Y<IMAGE->Rows; Y++)
         for(X=0; X<IMAGE->Cols; X ++)
            {
            *(IMAGEC->Data + X
            +(long)Y*IMAGE->Cols)=255-
            *(IMAGE->Data + X + (long)Y
            *IMAGE->Cols);
            }
         Erosion(IMAGE, M3, FILTER);
         Erosion(IMAGEC, M4, THICKEN);
         for(Y=N/2; Y<IMAGE->Rows-N/2; Y++)
           for(X=N/2; X<IMAGE->Cols-N/2; X++)
              *(FILTER->Data + X
              +(long)Y*IMAGE->Cols)=
              *(FILTER->Data + X +
              (long)Y*IMAGE->Cols) &
            *(THICKEN->Data + X
            +(long)Y*IMAGE->Cols);
         for(Y=N/2;Y<IMAGE->Rows-N/2;Y++)
            for(X=N/2;X<IMAGE->Cols-N/2;X++)
               {
               *(THICKEN->Data+X+(long)Y*R)=
               *(IMAGE->Data+X+(long)Y*R)|
               *(FILTER->Data +
               X+(long)Y*R);
               *(IMAGE->Data+X+(long)Y*R)=
               *(THICKEN->Data+X+(long)Y*R);
               }
         }
         stpflg++;
       }
       THICKEN->Rows=IMAGE->Rows;
       THICKEN->Cols=IMAGE->Cols;
```

SEE ALSO: Binary Hit-Miss, Closing, Opening, Erosion, Dilation, and Thinning Filters

CLASS: Morphological Filters

DESCRIPTION:

The binary thinning operation is used to reduce the geometric size of an object. Usually this operation is used iteratively until the desired image is obtained or until no further changes occur in the thinned image. The thinning operation is defined as

for i equal 1 to 8

$$A^{i+1} = A^{i} \cap [\text{HitMiss}(A^{i}, B^{i}, C^{i})]^{c}$$

$$\text{Thinning}(A) = A^{9}$$

where A^{1} is the original image, A^{9} is the thinned image, and B^{i} and C^{i} are the ith set of eight masks given by

B^{1}

0	0	0
1	1	1
1	1	1

C^{1}

1	1	1
0	0	0
0	0	0

B^{2}

1	1	1
1	1	1
0	0	0

C^{2}

0	0	0
0	0	0
1	1	1

B^{3}

0	1	1
0	1	1
0	1	1

C^{3}

1	0	0
1	0	0
1	0	0

B^{4}

1	1	0
1	1	0
1	1	0

C^{4}

0	0	1
0	0	1
0	0	1

B^5

1	0	0
1	1	0
1	1	1

C^5

0	1	1
0	0	1
0	0	0

B^6

1	1	1
0	1	1
0	0	1

C^6

0	0	0
1	0	0
1	1	0

B^7

1	1	1
1	1	0
1	0	0

C^7

0	0	0
0	0	1
0	1	1

B^8

0	0	1
0	1	1
1	1	1

C^8

1	1	0
1	0	0
0	0	0

EXAMPLES:

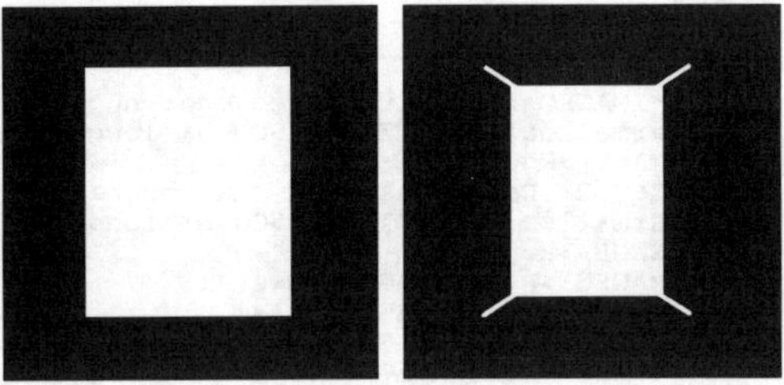

(a) ***(b)***

(a) The original image of a rectangle and ***(b)*** a partially thinned version.

ALGORITHM:

The program assumes that the original binary image is an IMAGE->Rows × IMAGE->Cols pixel image with a background graylevel value of 0 and an object graylevel value of 255 (object) and is stored in the structure IMAGE. Passed to the program in the variable ITERATION is the number of times the thinning operation is to be performed. Upon completion of the program, the thinned image is stored in the structure THINNED.

```
#define N 3

Thinned(struct Image *IMAGE,
      struct Image *THINNED, int ITERATION)
      {
      int X,Y,I,J,Z, M1[3][3][8], M4[3][3];
      int M2[3][3][8], stpflg=0, M3[3][3];
      long int R;
      struct Image *FILTER, *IMAGEC, A, A1;
      FILTER=&A;
      IMAGEC=&A1;
      /* Use these 2 lines for Non MS-DOS
      systems*/
      IMAGEC->Data=(unsigned char *)
      malloc((long)IMAGE->Cols*(long)
      IMAGE->Rows);
      FILTER->Data=(unsigned char *)
      malloc((long)IMAGE->Cols*(long)
      IMAGE->Rows);
      /*Use these 2 lines for MS-DOS systems
      */
      /*IMAGEC->Data=(unsigned char huge *)
      farmalloc((long)IMAGE->Cols*(long)
      IMAGE->Rows);
      FILTER->Data=(unsigned char huge *)
      farmalloc((long)IMAGE->Cols*(long)
      IMAGE->Rows);*/
      IMAGEC->Rows=IMAGE->Rows;
      IMAGEC->Cols=IMAGE->Cols;
      FILTER->Rows=IMAGE->Rows;
      FILTER->Cols=IMAGE->Cols;
      R=IMAGE->Cols;
      for(Z=0; Z<=7; Z++)
         for(J=0; J<=2; J++)
            for(I=0; I<=2; I++)
               M1[I][J][Z]=0;
```

```
M1[0][1][0]=1; M1[1][1][0]=1;
M1[2][1][0]=1;
M1[0][2][0]=1; M1[1][2][0]=1;
M1[2][2][0]=1;
M1[0][0][1]=1; M1[1][0][1]=1;
M1[2][0][1]=1;
M1[0][1][1]=1; M1[1][1][1]=1;
M1[2][1][1]=1;
M1[0][0][2]=1; M1[1][0][2]=1;
M1[0][1][2]=1;
M1[1][1][2]=1; M1[0][2][2]=1;
M1[1][2][2]=1;
M1[1][0][3]=1; M1[2][0][3]=1;
M1[1][1][3]=1;
M1[2][1][3]=1; M1[1][2][3]=1;
M1[2][2][3]=1;
M1[0][0][4]=1; M1[0][1][4]=1;
M1[1][1][4]=1;
M1[0][2][4]=1; M1[1][2][4]=1;
M1[2][2][4]=1;
M1[0][0][5]=1; M1[1][0][5]=1;
M1[2][0][5]=1;
M1[0][1][5]=1; M1[1][1][5]=1;
M1[0][2][5]=1;
M1[0][0][6]=1; M1[1][0][6]=1;
M1[2][0][6]=1;
M1[1][1][6]=1; M1[2][1][6]=1;
M1[2][2][6]=1;
M1[2][0][7]=1; M1[1][1][7]=1;
M1[2][1][7]=1;
M1[0][2][7]=1; M1[1][2][7]=1;
M1[2][2][7]=1;
for(Z=0; Z<=7; Z++)
   for(J=0; J<=2; J++)
      for(I=0; I<=2; I++)
      M2[I][J][Z]=1-M1[I][J][Z];
while(stpflg<ITERATION)
 {
 for(Z=0; Z<=7; Z++)
   {
   for(J=0; J<=2; J++)
      for(I=0; I<=2; I++)
         {
         M3[I][J]=M1[I][J][Z];
         M4[I][J]=M2[I][J][Z];
         }
for(Y=0; Y<IMAGE->Rows; Y++)
```

```
        for(X=0; X<IMAGE->Cols; X ++)
           {
           *(IMAGEC->Data + X
           +(long)Y*IMAGE->Cols)=255-
           *(IMAGE->Data + X + (long)Y
           *IMAGE->Cols);
           }
        Erosion(IMAGE, M3, FILTER);
        Erosion(IMAGEC, M4, THINNED);
        for(Y=N/2; Y<IMAGE->Rows-N/2; Y++)
          for(X=N/2; X<IMAGE->Cols-N/2; X++)
            *(FILTER->Data + X
            +(long)Y*IMAGE->Cols)=
            *(FILTER->Data + X +
            (long)Y*IMAGE->Cols) &
           *(THINNED->Data + X
           +(long)Y*IMAGE->Cols);
        for(Y=N/2;Y<IMAGE->Rows-N/2;Y++)
           for(X=N/2;X<IMAGE->Cols-N/2;X++)
              {
              *(THINNED->Data+X+(long)Y*R)=
              *(IMAGE->Data+X+(long)Y*R)&
              (255-*(FILTER->Data +
              X+(long)Y*R));
              *(IMAGE->Data+X+(long)Y*R)=
              *(THINNED->Data+X+(long)Y*R);
              }
        }
        stpflg++;
      }
      THINNED->Rows=IMAGE->Rows;
      THINNED->Cols=IMAGE->Cols;
     }
```

SEE ALSO: Binary Hit-Miss, Closing, Opening, Erosion, Dilation, and Thickening Filters

CLASS: Segmentation

DESCRIPTION:

Thresholding is used in image processing to separate an object's pixels from the background pixels. Thresholding converts a multi-graylevel image into a binary image containing two graylevel values. The threshold operation is defined as

$$g(x, y) = \begin{cases} G_o & \text{if } f(x, y) > T \\ G_b & \text{if } f(x, y) \leq T \end{cases},$$

where f(x, y) is the original image, g(x, y) is the binarized image, T is the threshold value, G_o is the object graylevel value after the thresholding operation, and G_b is the background gravlevel value after the thresholding operation.

EXAMPLE:

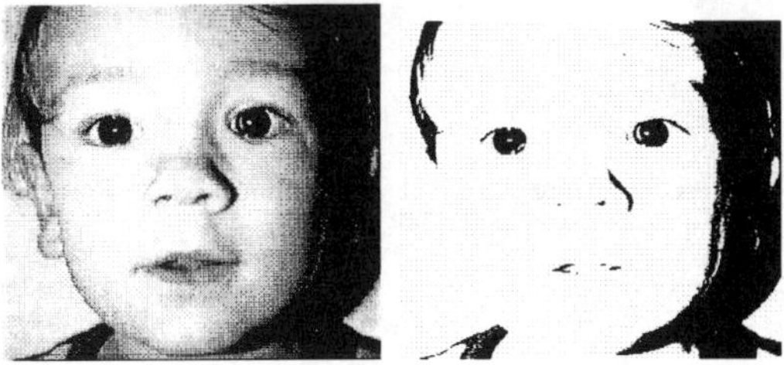

(a) *(b)*

(a) The original image and ***(b)*** the thresholded image with T = 70.

ALGORITHM:

The program assumes that the original image is a 256 graylevel × IMAGE->Rows × IMAGE->Cols pixel image stored in the structure IMAGE. The program then compares each pixel of the image against the threshold parameter THRES. If a pixel's graylevel value is greater than the

THRES parameter it is set to the graylevel given by the GO parameter; otherwise it is set to the graylevel value given by the GB parameter. Upon completion of the program, the resulting binary image is stored in the structure IMAGE1.

```
Threshold(struct Image *IMAGE, struct Image
*IMAGE1, int THRES)
      {
      int X, Y, GO, GB;
      GO=255;
      GB=0;
      for(Y=0; Y<=IMAGE->Rows; Y++)
         {
         for(X=0; X<=IMAGE->Cols; X++)
            {
            if(*(IMAGE->Data+X+(long)Y*
            IMAGE->Cols) > THRES)
                  *(IMAGE1->Data+X+(long)Y*
            IMAGE->Cols)= GO;
            else
                  *(IMAGE1->Data+X+(long)Y*
            IMAGE->Cols)= GB;
            }
         }
      }
```

SEE ALSO: Multi-graylevel Thresholding, Point Detector, Line Detectors, and Optimum Thresholding

CLASS: Morphological Filters

DESCRIPTION:

Morphological top-hat of an image is defined as the difference between the original image and the graylevel opened image. The top-hat filter is used to enhance low contrast high spatial frequency features within an image.

$$\text{TopHat}(A, B) = A - \text{open}(A, B)$$

EXAMPLES:

(a)

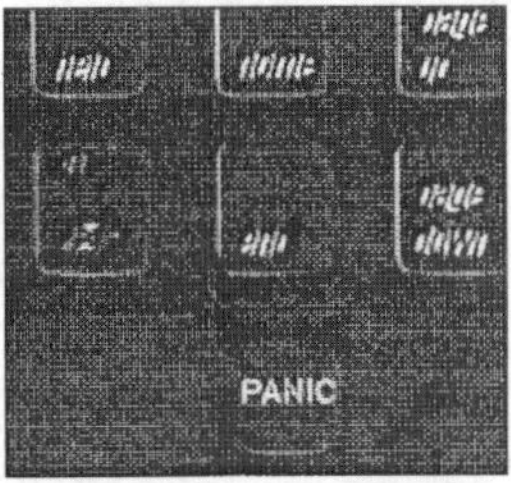

(b)

(a) The original image and ***(b)*** the top-hat image using an all zero 7×7 mask.

ALGORITHM:

The program assumes that the original image is a 256 graylevel × IMAGE->Rows × IMAGE->Cols pixel image stored in the structure IMAGE. The N × N structuring function is stored in the array MASK[][]. Upon completion of the program, the top-hat image is stored in the structure FILTER. The graylevel opening filter used by the algorithm can be found under graylevel opening.

```
#define N 5

TopHat(struct Image *IMAGE, int
   MASK[][N],struct Image *FILTER)
      {
      int X, Y, B;
      struct Image *TEMP, A;
      TEMP=&A;
      /* Use this line for Non MS-DOS
      systems*/
      TEMP->Data=(unsigned char *)
      malloc((long)IMAGE->Cols*(long)
      IMAGE->Rows);
      /*Use this line for MS-DOS systems
      */
      /*TEMP->Data=(unsigned char huge *)
      farmalloc((long)IMAGE->Cols*(long)
      IMAGE->Rows);*/
      TEMP->Rows=IMAGE->Rows;
      TEMP->Cols=IMAGE->Cols;
      for(Y=0; Y<IMAGE->Rows; Y++)
         {
         for(X=0; X<IMAGE->Cols; X++)
            {
            *(TEMP->Data+X+(long)Y
            *IMAGE->Cols)=
            *(IMAGE->Data+X+(long)Y*
            IMAGE->Cols);
            }
         }
      OpenGray(IMAGE, MASK, FILTER);
      for(Y=0; Y<IMAGE->Rows; Y++)
         {
         for(X=0; X<IMAGE->Cols; X++)
            {
            B=*(TEMP->Data+X+(long)Y*
            IMAGE->Cols)-
```

```
        *(FILTER->Data+X+(long)
        Y*IMAGE->Cols);
        if(B<0)
        B=0;
        *(FILTER->Data+X+(long)Y
        *IMAGE->Cols)=B;
        }
    }
  FILTER->Rows=IMAGE->Rows;
  FILTER->Cols=IMAGE->Cols;
  }
```

SEE ALSO: Graylevel Opening, Closing, Erosion, and Dilation Filters

Transforms

DESCRIPTION:

A large area of image processing is two-dimensional image transforms. These transforms are used to enhance, restore, and compress images. Most important of the transforms is the discrete Fourier transform. This transform decomposes an image into a set of sine and cosine functions at different frequencies. The components generated from the discrete Fourier transform are then used to compute the magnitude and phase spectrums of the image. Spatial frequency filtering of an image involves the modification of these spectral components and is typically used to enhance an image. Other transforms such as the discrete cosine, Walsh, and Hadamard transforms are used in image compression to reduce the storage size of an image.

CLASS MEMBERSHIP:

- Discrete Cosine Transform
- Discrete Fourier Transform
- Discrete Fourier Transform Properties
- Hadamard Transform
- Hartley Transform
- Hough Transform
- Slant Transform
- Walsh Transform

SEE ALSO: Coding & Compression.

CLASS: Color Image Processing

DESCRIPTION:

What separates a true-color display system from a pseudocolor display system is that there are three input image memories, one for each primary color. Figure ***a*** is the block diagram for a true-color display system. Each image memory contains the appropriate color intensity value for each of the three primary colors for every pixel within the image. In this way, true-color viewing of a color image is possible. Each color image pixel can have one out of 16 million possible colors. With computer memory devices becoming cheaper, more imaging systems are becoming available that support true-color capability, which is also known as a 24-bit color system.

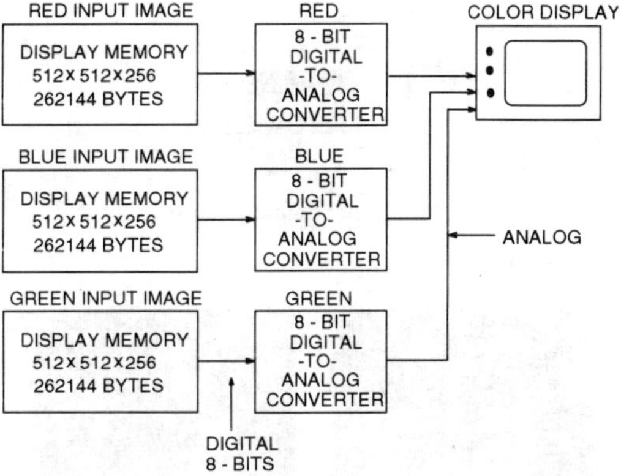

(a) True-color display system

SEE ALSO: RGB, HSI and YIQ Color Models, C.I.E. Color Chart, and Pseudocolor Display

CLASS: Noise

DESCRIPTION:

Uniform noise is commonly used to degrade images in the evaluation of image processing algorithms. The uniform noise histogram is defined as

$$h_i = \begin{cases} \dfrac{1}{b-a} & \text{for } a \le G_i \le b \\ 0 & \text{elsewhere} \end{cases}$$

where G_i is the ith graylevel value of the image, and the parameters a and b are the minimum and maximum graylevel values of the uniform noise.

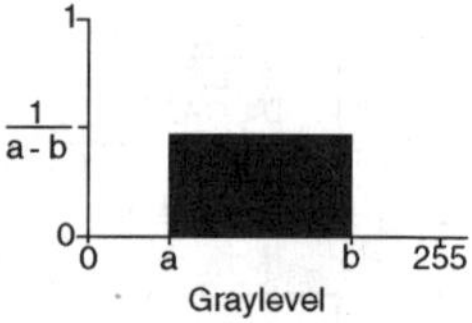

A histogram of uniform noise.

EXAMPLES:

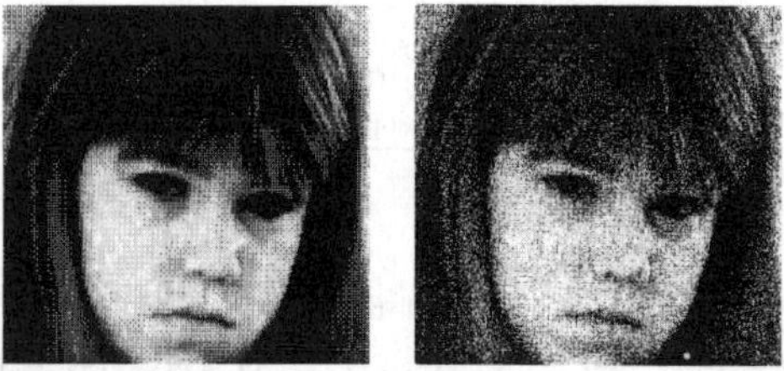

(a) *(b)*

(a) The original image and ***(b)*** the (additive) uniform noise degraded image with a variance = 800 and a Mean = 0.

ALGORITHM:

The program generates a uniform noise image of 256 graylevel × IMAGE->Rows × IMAGE->Cols pixel image stored in the structure IMAGE. The program assumes the function rand() generates a uniform random number in the range of 0 to 32767. Both the desired mean and variance are passed to the program upon execution. If the noise graylevel value generated exceeds the 256 graylevel range, the noise graylevel value is truncated to either 0 or 255.

```
Uniform(struct Image *IMAGE,
      float VAR, float MEAN)
      {
      int X, Y;
      float NOISE;
      for(Y=0; Y<IMAGE->Rows; Y++)
         for(X=0; X<IMAGE->Cols; X++)
            {
            NOISE = sqrt((double)VAR)
            *1.057192E-4 *(float)rand()
            + MEAN - sqrt((double)VAR)
            *1.7320508;
            if(NOISE > 255)
                 NOISE = 255;
            if(NOISE < 0)
                 NOISE = 0;
            *(IMAGE->Data +X +(long)Y*
            IMAGE->Cols) = (unsigned
            char)(NOISE +.5);
            }
      }
```

SEE ALSO: Rayleigh, Gaussian, Negative Exponential, Salt and Pepper and Gamma Noises

CLASS: Transform

DESCRIPTION:

A transform that decomposes an image into a set of square waves is known as the Walsh transform. It is typically used in image compression. Another transform similar to the Walsh transform is the Hadamard transform. The two-dimensional Walsh transform of a $N \times N$ image is defined as

$$F(n, m) = \frac{1}{N^2} \sum_{Y=0}^{N-1} \sum_{X=0}^{N-1} f(X, Y) \cdot \prod_{i=0}^{q-1} (-1)^{b_i(X) b_{q-1-i}(n) + b_i(Y) b_{q-1-i}(m)}$$

and its inverse

$$f(X, Y) = \sum_{m=0}^{N-1} \sum_{m=0}^{N-1} F(n, m) \cdot \prod_{i=0}^{q-1} (-1)^{b_i(X) b_{q-1-i}(n) + b_i(Y) b_{q-1-i}(m)},$$

where q is the total number of bits, i.e., $N = 2^q$ and $b_i(x)$ is the ith bit of the binary representation of the number x. For example, if the total number of bits (q) equals 4 (hence N = 16) and X equals 7 (0111), then $b_0(7) = 1$, $b_1(7) = 1$, $b_2(7) = 1$, and $b_3(7) = 0$.

ALGORITHM:

The program assumes that the original image is a floating point square image of size IMAGE->Rows stored in the floating point structure IMAGE. Passed to the program is the variable *dir* which is used to determine if a forward (dir = 1) or if an inverse (dir = 0) Walsh transform is to be computed. The program first sets the unsigned character pointer *(B + i + x · q) of size q × IMAGE->Rows to the 1/0 bit representation of the number x using the *bitrep* subroutine. The index x is used to access the number and the index i is used to access the bit within the number. The least significant bit of the number x corresponds to the index i equal to zero. The program then computes the Walsh transform in the x direction followed by the Walsh transform in the y direction. Upon completion of the program, the Walsh components are returned in the floating point structure IMAGE1.

```
Walsh(struct  Image  *IMAGE,  struct  Image
*IMAGE1, int dir)
      {
      int X, Y, n, m, num, I, q;
      int prod, A, temp;
      unsigned char *B;
      float K0, sum;
      num=IMAGE->Rows;
      q=(int)(log((double)
      IMAGE->Rows)/log(2.0)+.5);
      B=malloc(num*q);
      bitrep(B,q,num);
      if(dir==1)
        {
        K0=num*num;
        for(m=0; m<num; m++)
          {
          for(n=0; n<num; n++)
            {
            sum=0;
            for(Y=0; Y<num; Y++)
               {
               for(X=0; X<num; X++)
                  {
                  prod=1;
                  for(I=0;I<=q-1;I++)
                     {
                     A=
                     *(B+I+X*q) *
                     *(B+q-1-I+n*q) +
                     *(B+I+Y*q) *
                     *(B+q-1-I+m*q);
                     if((A/2)*2==A)
                        temp=1;
                     else
```

```
                    temp=-1;
                  prod=prod*temp;
                  }
                sum=sum+ *(IMAGE->Data
                +X+(long)Y*
                IMAGE->Rows)*prod;
                }
              }
          *(IMAGE1->Data+n+(long)m*
          IMAGE->Rows)=sum;
          }
        }
        for(X=0; X<num; X++)
            for(Y=0; Y<num; Y++)
                *(IMAGE1->Data+X+(long)Y*
                IMAGE->Rows)=
                *(IMAGE1->Data+X+(long)Y*
                IMAGE->Rows)/K0;
      }
    if(dir==-1)
      {
      for(Y=0; Y<num; Y++)
        {
        for(X=0; X<num; X++)
          {
          sum=0;
          for(m=0; m<num; m++)
              {
              for(n=0; n<num; n++)
                  {
                  prod=1;
                  for(I=0;I<=q-1;I++)
                      {
                      A=
                      *(B+I+X*q) *
                      *(B+q-1-I+n*q) +
                      *(B+I+Y*q) *
                      *(B+q-1-I+m*q);
                      if((A/2)*2==A)
                         temp=1;
                      else
                         temp=-1;
                      prod=prod*temp;
                      }
                  sum=sum+ *(IMAGE->Data
                  +n+(long)m*
                  IMAGE->Rows)*prod;
```

```
                }
            }
          *(IMAGE1->Data+X+(long)Y*
          IMAGE->Rows)=sum;
          }
        }
      }
    }

bitrep(unsigned char *b, int q, int num)
     {
     int x,i, bit;
     for(x=0;x<num;x++)
        {
        bit=1;
        for(i=0;i<q;i++)
           {
           *(b+i+x*q)= (x&bit)/bit;
           bit=bit<<1;
           }
        }
     }
```

SEE ALSO: Fourier Transform Properties, the Hadamard Transform, the Discrete Cosine Transform, and the Discrete Fourier Transform

CLASS: Graphics Algorithms

DESCRIPTION:

Warping is the use of image mapping functions to produce geometric distortions in images. The most common transformation is based on the affine projection, given by the following set of equations:

$$X' = \frac{\mathbf{aX + bY + c}}{\mathbf{ix + jY + 1}}$$

$$Y' = \frac{\mathbf{dX + eY + f}}{\mathbf{ix + jY + 1}}$$

X, Y are the old coordinates and X', Y' the new. The coefficients, **a**, **b**, **c**, **d**, **e**, **f**, **i**, and **j** are determined from a set of four control points that correspond to the congruency desired between the two images and the selected template. The new coordinates do not all map to integer pixel coordinates, thus it is necessary to interpolate the final result in some way. The simplest method for doing this is using a bilinear interpolation scheme that examines the nearest neighbor values and averages between them.

EXAMPLE:

ALGORITHM:

The algorithm given warps <u>two</u> rectangular coordinates from the input image, **Img**, into any <u>four</u> coordinates of the output image, **Out**. The input rectangle is supplied by the integer coordinate pairs (**sx**, **sy**) and (**ex**, **ey**). The four warp-to coordinate pairs are supplied as pointers to two integer arrays, **Wx** and **Wy**. The routine assumes that the coordinates are consistent and that the output image will accommodate the warp result. Multiple calls to this routine can effect complex warping schemes on an image; however, no attempt has been made to optimize the process in terms of speed or interpolative smoothness. The warp algorithm may also be used for interpolated zoom and dezoom as well as rotation and translation operations.

```
/* Basic Warp Algorithm using
   bilinear interpolation */

warp(struct Image *Img, struct Image *Out,
int sx,int sy,int ex,int ey,int *Wx,int *Wy)
{

   float a,b,c,d,e,f,i,j,x,y,destX,destY;
   int dy;

   /* Set up Warp coefficients */
   a = (float)(-(*Wx) + *(Wx+1))/(ey-sy);
   b = (float)(-(*Wx) + *(Wx+3))/(ex-sx);
   c = (float)((*Wx) - *(Wx+1) + *(Wx+2) -
                 *(Wx+3))/((ey-sy)*(ex-sx));
   d = (float)(*Wx);
   e = (float)(-(*Wy) + *(Wy+1))/(ex-sx);
   f = (float)(-(*Wy) + *(Wy+3))/(ey-sy);
   i = (float)((*Wy)- *(Wy+1)+ *(Wy+2)-
                *(Wy+3))/((ey-sy)*(ex-sx));
   j = (float)(*Wy);

   /* warp */
   for(y=sy; y<ey; y+=0.5){
      dy = (int)y+0.5;
      for(x=sx; x<ex; x+=0.5){
         destX = a*x + b*y + c*x*y + d;
         destY = e*x + f*y + i*x*y + j;
         *(Out->Data + (Out->Cols *
(int)(destY+0.5)) + (int)destX) =
      *(Img->Data +
      (long)(y*Img->Cols)+(long)(x+.05));
      }
   }
} /*end warp*/
```

SEE ALSO: Morphing, Zooming, Rotation

CLASS: Spatial Filters

DESCRIPTION:

A weighted mean filter is used so that different weights can be used for each pixel in the average calculation. The weighted mean filter is defined as the weighted average of all pixels within a local region of an image. Weights for each pixel and which pixels are included in the averaging operation are specified by an input mask. The larger the filtering mask becomes the more predominant the blurring becomes and less high spatial detail remains in the filtered image. The definition of a weighted mean filter in terms of an image A is

$$\text{Mean}\,(A) = \frac{\sum\limits_{(i,\,j)\,\in\,M} w_{i,\,j} \cdot A(x+i,\; y+j)}{\sum\limits_{(i,\,j)\,\in\,M} w_{i,\,j}}\;,$$

where the coordinate x + i, y + j is defined over the image A and the coordinate i, j is defined over the mask M. The weight $w_{i,\,j}$ is the weight of the mask associated with the pixel located at the coordinate (i + x, j + y) within the image.

EXAMPLE:

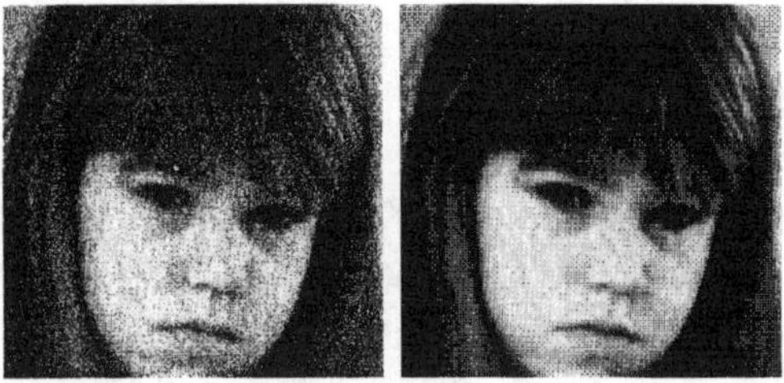

(a) ***(b)***

(a) A Gaussian noise corrupted image and ***(b)*** the weighted mean filtered image using the 5 × 5 mask given in the text.

ALGORITHM:

The program assumes that the original image is a 256 graylevel × IMAGE->Rows × IMAGE->Cols pixel image stored in the structure IMAGE. The program then performs a N × N weighted mean filter on the image with the weights passed to the program in the floating point array MASK[][]. The size of the filtering operation is determined by the #define NUM 7 statement and should be set to an odd number. Upon completion of the program, the filtered image is stored in the structure IMAGE1.

Sample 5 × 5 mask

1	1	1	1	1
1	2	2	2	1
1	2	3	2	1
1	2	2	2	1
1	1	1	1	1

```
#define NUM 7

WeightedMean(struct Image *IMAGE, struct
Image *IMAGE1,  float MASK[][NUM])
      {
      int X, Y;
      int I, J;
      int N;
      float SUM, SUMW;
      N=NUM;
      for(Y=N/2; Y<IMAGE->Rows-N/2; Y++)
         {
         for(X=N/2; X<IMAGE->Cols-N/2; X++)
            {
            SUMW=0.0; SUM=0.0;
            for(J=-N/2; J<=N/2; J++)
               {
               for(I=-N/2; I<=N/2; I++)
                  {
                  SUM=SUM + *(IMAGE->Data+X
                  +I+(long)(Y+J)*IMAGE->Cols)
                  * MASK[I+N/2][J+N/2];
```

```
                SUMW=SUMW +
                MASK[I+N/2][J+N/2];
                }
            }
            *(IMAGE1->Data+X+(long)Y
            *IMAGE->Cols)=
            (unsigned char)(SUM/SUMW +.5);
        }
    }
}
```

SEE ALSO: Median, Minimum, Maximum, and other Nonlinear Filters

CLASS: Nonlinear Filters

DESCRIPTION:

A weighted median differs from a median filter in that specified pixels within a local neighborhood are repeated a given number of times in the computation of the median value. The weighted median is defined as

$$\text{Median}(A) = \text{Median}[\text{Repeat } Q_{i,j} \text{ times}\{A((x+i,\ y+j)\}],$$

where the coordinate x + i, y + j is defined over the image A and the coordinate i, j is defined over the mask M. The mask M determine which pixels are to be included in the median calculation and the values of the mask $Q_{i,j}$ determine how many times each pixel within the mask is to be repeated in the median calculation. For example, the following mask, when used with the weighted median filter, will preserve one pixel wide horizontal lines.

1	1	1
3	3	3
1	1	1

EXAMPLE:

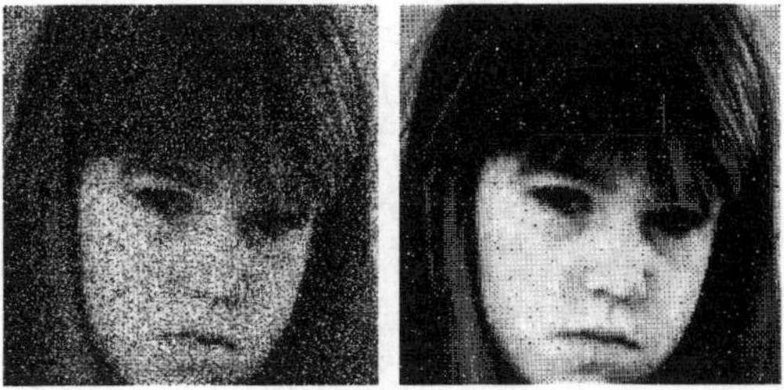

(a) *(b)*

(a) The original salt and pepper noise corrupted image and ***(b)*** the weighted median filtered image using the 3 × 3 mask given in the text above.

ALGORITHM:

The program assumes that the original image is a 256 graylevel × IMAGE->Rows × IMAGE->Cols pixel image stored in the structure IMAGE. The program then performs a N × N weighted median filter on the image using the mask passed to the program. The numbers in the mask determine the number of times each pixel is repeated in the median calculation. The only requirement is that the sum of all the numbers in the mask must add up to an odd number. The size of the filtering operation is determined by the constant N and should also be set to an odd number. The total number of values included in the median calculation must be less than 150. Upon completion of the program, the filtered image is stored in the structure IMAGE1.

#define N 7

```
WeightedMedian(struct Image *IMAGE, struct
Image *IMAGE1,  int MASK[][N])
      {
      int X, Y, I, J, M, Z;
      int AR[150], A;
      for(Y=N/2; Y<IMAGE->Rows-N/2; Y++)
         for(X=N/2; X<IMAGE->Cols-N/2; X++)
            {
            Z=0;
            for(J=-N/2; J<=N/2; J++)
               for(I=-N/2; I<=N/2; I++)
                  {
                  for(M=1;M<=MASK[(int)N/2
                  +I][(int)N/2+J]; M++)
                     {
                     AR[Z]=*(IMAGE->Data+X
                     +I+(long)(Y+J)
                     *IMAGE->Cols);
                     Z++;
                     }
                  }
            for(J=1; J<=Z-1;J++)
               {
               A = AR[J];
               I=J-1;
               while(I>=0 && AR[I] >A)
                  {
                  AR[I+1]=AR[I];
```

```
                I=I-1;
                }
                AR[I+1]=A;
            }
            *(IMAGE1->Data+X +(long)Y
            *IMAGE->Cols)=AR[Z/2];
        }
    }
```

SEE ALSO: Arithmetic Mean, Minimum, Maximum, Median, and other Nonlinear Filters

CLASS: Spatial Frequency Filters

DESCRIPTION:

The *Wiener Filter*, also known as the Least Mean Square filter, is given by the following expression:

$$\hat{F}(u,v) = \left[\frac{H(u,v)^*}{|H(u,v)|^2 + [S_n(u,v)/S_f(u,v)]}\right] G(u,v)$$

H(u,v) is the degradation function (* indicates complex conjugate) and G(u,v) is the degraded image. The functions $S_f(u,v)$ and $S_n(u,v)$ are the power spectra of the original image (prior to degradation) and the noise. If the noise is zero, the filter reduces to the inverse filter. If the power spectral densities are unknown, they may be replaced by a constant scalar. In this case, the evaluation may be performed incrementally while changing the constant until a satisfactory result is obtained.

The example shows the keyboard picture corrupted by noise and the Wiener filtered result to the right. This picture was prepared from an estimate of the power densities and not an exact evaluation.

EXAMPLE:

ALGORITHM:

The algorithm for the Wiener filter is identical to that of the *parametric* wiener filter, with the gamma term set equal to 1.

SEE ALSO: Inverse Filter, Least Squares Filter, Wiener Filter (parametric)

CLASS: Spatial Frequency Filters

DESCRIPTION:

The *ParametricWiener Filter* is given by the following expression:

$$\hat{F}(u,v) = \left[\frac{H(u,v)^{*}}{|H(u,v)|^{2} + \gamma\left[S_{n}(u,v)/S_{f}(u,v)\right]} \right] G(u,v)$$

H(u,v) is the degradation function and G(u,v) is the degraded image. The functions $S_f(u,v)$ and $S_n(u,v)$ are the power spectra of the original image (prior to degradation) and the noise. If the noise is zero, the filter reduces to the inverse filter.

The γ is the Wiener parameter and is adjusted so that the least squared error constraint is satisfied. The filter can be shown to be optimal when $\gamma = 1$. The parameter can be adjusted and thus affect the outcome of the filter by giving more or less weight to the effect of the power spectrum terms. In this case, the parameter is adjusted incrementally after each filter pass.

If the power spectral densities are unknown, they may be replaced by a constant scalar. Once again, the evaluation may be incremental until a satisfactory result is obtained.

The example shows thc keyboard picture corrupted by noise and the Wiener filtered result to the right. This picture was prepared from an estimate of the power densities and not an exact evaluation.

EXAMPLE:

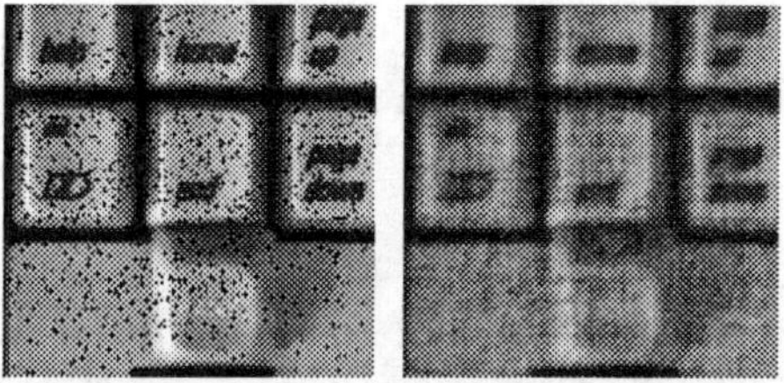

ALGORITHM:

The algorithm given computes the parametric Wiener filter on the Fourier transform of a degraded image, **Guv**, with noise image spectra **N**, degradation function **Huv**, and original image **Img**. The computation is *in place* so that the filtered version of the input is returned in the original image variable. The original and noise images are either estimations from some predictive function or *ad hoc* approximations. If the noise image is zero, the process reduces to the inverse filter.

The Wiener parameter gamma is passed to the algorithm as **g**. If this parameter is 1.0, the filter is non-parametric. Methods exist in the literature to derive the parameter value, however, it is sometimes determined from trial and error.

```
void p_wiener_filter(struct Image *Img,
    struct Image *Huv, struct Image *N,
    struct Image *Guv, float g)
{
   extern void cxdv(float a,float b,float c,
          float d,float *R,float *I),
   cxml(float a,float b,float c,
          float d,float *R,float *I);

   float *In, *No, *H, *G;
   long i,sz;
   float Nr,Ni,Dr,Di,Hsr,Hsi;

   In = (float *)(Img->Data);
   H  = (float *)(Huv->Data);
   No = (float *)(N->Data);
   G  = (float *)(Guv->Data);

   sz = Img->Rows * Img->Cols;

   for(i=0; i<sz;i+=2){
      /* noise spectral density, computed
         in place  with wiener gamma.
         IF:
               gamma == 1, wiener filter
         ELSE:
               parametric wiener
      */
      cxml(g*(*(No+i)),g*(*(No+i+1)),*No+i,
         -1.0*(*(No+i+1)),No+i,No+i+1);
```

```
      /* image spectral density */
      cxml(*(In+i),*(In+i+1),*(In+i),
           -1.0*(*(In+i+1)),&Dr,&Di);

      /* denominator spectral density term */
      cxdv(*(No+i),*(No+i+1),Dr,Di,&Dr,&Di);

      /* degradation power spectrum */
      cxml(*(H+i),*(H+i+1),*(H+i),
           -1.0*(*(H+i+1)),&Hsr,&Hsi);

      /* numerator term */
      cxml(*(H+i),-1.0*(*(H+i+1)),*(G+i),
           *(G+i+1),&Nr,&Ni);

      /* final calculation */
      cxdv(Nr,Ni,Hsr+Dr,Hsi+Di,In+i,In+i+1);
   }
}

/* Complex divide */
void cxdv(float a,float b,float c,
          float d,float *R,float *I)
{
   float denom;

   denom = (c*c)+(d*d);
   if(denom == 0.0){
      *R = 0;
      *I = 0;
   }
   *R = ((a*c)+(b*d))/denom;
   *I = ((b*c)-(a*d))/denom;
}

/* Complex multiply */
void cxml(float a,float b,float c,
          float d,float *R,float *I)
{
   *R = (a*c) - (b*d);
   *I = (a*d) - (b*c);
}
```

SEE ALSO: Inverse Filter, Least Squares Filter, Wiener Filter

CLASS: Color Image Processing

DESCRIPTION:

The YIQ color system is based upon the color television standard. It divides a color into three components: its luminance (Y), an in phase (I), and a quadrature phase (Q). The hue and saturation information of a color is contained in the I and Q color components. The YIQ components in terms of the three primary colors red (R), green (G), blue (B) are

$$Y = 0.30R + 0.59G + .11B$$

$$I = 0.28G + 0.59R - 0.32B$$

$$Q = -0.53G + 0.21R + 0.31B$$

Figure ***a*** shows a two-dimensional plot of the I and Q components. The angle θ presents the tint of the color while the length of the color vector τ from the origin represents the purity or saturation of the color. The larger the vector τ is, the purer the color is. At the origin, the color only contains luminance information. We call this color a black and white color.

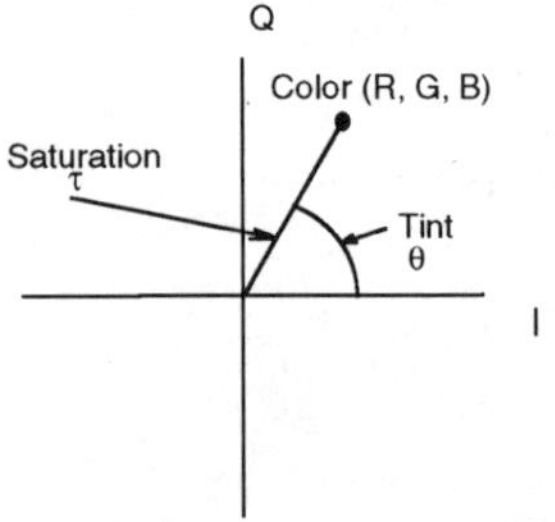

(a) A plot of the I and Q color components

SEE ALSO: HSI, RGB and YIQ Color Models, C.I.E. Color Chart and Pseudocolor

CLASS: Nonlinear Filters

DESCRIPTION:

The Y_p mean filter is member of a set of nonlinear mean filters which are better at removing Gaussian type noise and preserving edge features than the arithmetic mean filter. The Y_p mean filter is very good at removing positive outliers for negative values of P and negative outliers for positive values of P. If all the pixels included in the calculation of the harmonic mean are zero, the output of the Y_p filter will also be zero. The definition of the Y_p mean filter is

$$Y_p \operatorname{Mean}(A) = \left\{ \sum_{(i,j) \in M} \frac{A(x+i,\ y+j)^P}{N} \right\}^{1/P},$$

where the coordinate x + i, y + j is defined over the image A and the coordinate i, j is defined over the mask M. The mask M determines which pixels are to be included in the Y_p mean calculation and the value of N is the number of pixels used. The parameter P chooses the order of the filter.

EXAMPLE:

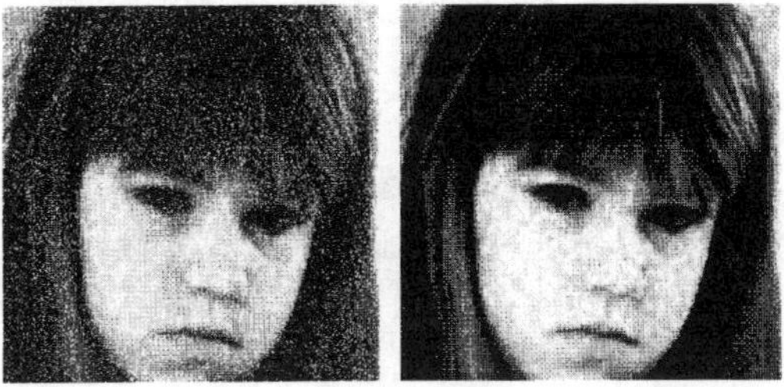

(a) ***(b)***

(a) The 10% positive outlier corrupted image and ***(b)*** the Y_p mean filtered image using a 3 × 3 square mask and P = -2.

ALGORITHM:

The program assumes that the original image is a 256 graylevel × IMAGE->Rows × IMAGE->Cols pixel image stored in the structure IMAGE. The program computes the Y_p filter over a set of pixels contained within a square N × N region of the image centered at the pixel X, Y. The size of the filtering operation is determined by the variable N and should be set to an odd number and be less than 12. Upon completion of the program, the filtered image is stored in the structure IMAGE1.

```
YpMean(struct Image *IMAGE, int P, struct
Image *IMAGE1)
      {
      int X, Y;
      int I, J, Z;
      int N, AR[121], A;
      float SUM;
      N=5;
      for(Y=N/2; Y<IMAGE->Rows-N/2; Y++)
         for(X=N/2; X<IMAGE->Cols-N/2; X++)
            {
            Z=0;
            for(J=-N/2; J<=N/2; J++)
               for(I=-N/2; I<=N/2; I++)
                  {
                  AR[Z]=*(IMAGE->Data+X
                  +I+(long)(Y+J)
                  *IMAGE->Cols);
                  Z++;
                  }
            Z=0;
            SUM=0.0;
            for(J=0; J<=N*N-1;J++)
               {
               if(AR[J]==0 && P<0)
                  Z=1;
               else
                  SUM=SUM+pow((double)AR[J],
                  (double)P);
               }
            if(Z==1)
               *(IMAGE1->Data+X +(long)Y
               *IMAGE->Cols)=0;
            else
               {
```

```
            if(SUM==0.0)
               A=0.0;
            else
               A=(int)pow((double)SUM/
               (double)(N*N),(double)(1.0
               /P));
            if(A >255)
                 A = 255;
            *(IMAGE1->Data+X +(long)Y
            *IMAGE->Cols)=A;
          }
      }
}
```

SEE ALSO: Geometric, Contra-Harmonic, Harmonic and Arithmetic Mean Filters, Median, and other Nonlinear Filters

CLASS: Graphics Algorithms

DESCRIPTION:

Zooming (Dezooming) algorithms magnify (or minify) images. Since the size and number of pixels ultimately determines the size of a picture, zooming creates pixels, and dezooming deletes pixels. In the simplest realization, we can zoom an image 200% by simply duplicating each pixel into a four-pixel neighborhood and dezoom to 50% by removing three pixels from each four-pixel neighborhood (subsample). This is illustrated in the graphic below:

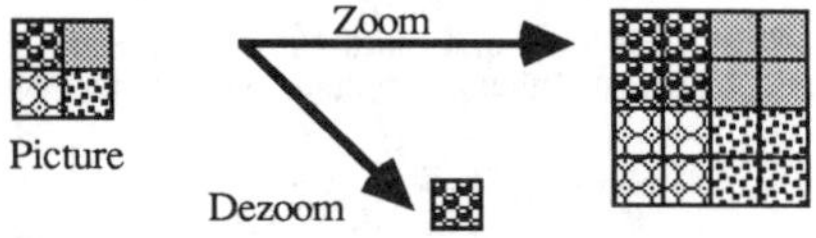

This method is simple to implement but requires that the final size be an integer multiple of two. Also, zooming reduces the sampling rate of the picture and causes blocking effects. Dezooming removes information from the original picture that cannot be recovered from the minimized image. Interpolation methods minimize the blocking effect problems (and allow for fractional zoom/dezoom). In these algorithms, an estimate is made of what the graylevels in the transformed pixel neighborhood should be. The basis of this estimate yields various interpolative algorithms. The example picture shows the result of a noninteger zoom (125%) and dezoom (75%) using an interpolation algorithm. Interpolative zooming may also be accomplished using a *warping* algorithm, where the image is warped to a larger (smaller) proportional shape.

EXAMPLE:

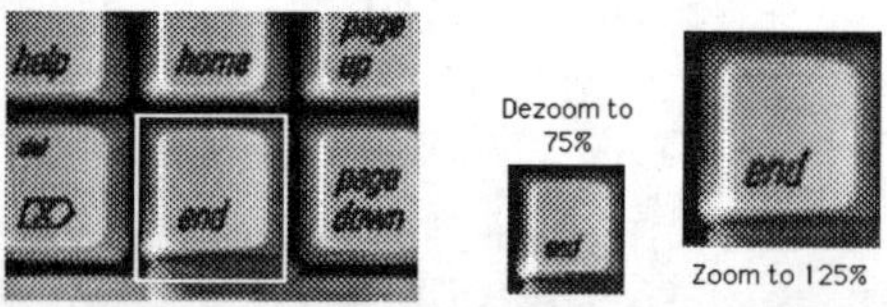

ALGORITHM:

The **zoom** routine accepts two pointers to unsigned character image data structures, **In** and **Out**. The start coordinates, **x** and **y**, and the horizontal (**hor**) and vertical (**ver**) sizes of the region to be operated on are also passed. The operation flag **zd** is set to nonzero for zooming and zero for dezooming. The macro **idx** is used for simplified access to the pixels of the image by coordinates.

```
#define idx(Im,i,j) \
        *(Im->Data + (i)*Im->Cols + (j))

/* Zoom/Dezoom */
zoom(struct Image *In, struct Image *Out,
     int x, int y, int hor,int ver,
     unsigned char zd)
{
   int i,j,m=0, n=0;

   if(zd){ /* Zoom */
      for(i=y;i<ver;++i){
         for(j=x;j<hor;++j){
            idx(Out,m,n)       =
            idx(Out,m+1,n)     =
            idx(Out,m,n+1)     =
            idx(Out,m+1,n+1)  = idx(In,j,i);
            n += 2;
         }
         m += 2;
         n = 0;
      }
   } else {   /* DeZoom */
      for(i=y;i<ver;i+=2){
         for(j=x;j<hor;j+=2){
            idx(Out,m,n) = idx(In,i,j);
            ++n;
         }
         ++m;
         n = 0;
      }
   }
}
```

SEE ALSO: Warping

```
/* Image.h include file

      Handbook of
      Image Processing Algorithms
      in C

   Harley R. Myler
   Arthur R. Weeks          */

struct Image {
   /* # of rows in image */
   int Rows;
   /* # of columns in image */
   int Cols;
   /* Pointer to image data */
   unsigned char *Data;
   /* type of image */
   unsigned char Type;
};

/* Image Types */

/* unsigned character image */
#define BASIC   0

/* unsigned integer image */
#define UINT    1

/* float image */
#define REAL    2

/* complex image float real,imaginary */
#define CMPLX   4
```

NOTE: for DOS applications, the Data pointer should be specified **huge**.

This appendix contains the complete source for a program to compute the Laplacian on an image using the routines contained in this handbook. You are free to copy it and use it as the shell, or template, for your own image processing programs. We also describe various image input and output routines.

The program uses dynamic allocation of memory for the image arrays. Images used are 256 × 256 pixels and the *Data* pointer of the Image structures *In* and *Out* are dynamically assigned to the 65,536 byte data blocks returned by the *calloc* function. *Calloc* is a common memory allocation routine available to C compilers. An include file that contains the function declaration is often required; this example uses *memory.h*. One must use caution when allocating large arrays on small systems. *Calloc* requires two arguments, the number of blocks desired and the size of the blocks. We found it convenient to specify the number of rows as the first argument and the number of columns as the second. If unusual results or crashes occur, it is most often the allocation, or misallocation, of memory.

For MSDOS™ systems, the data pointers should be initialized as *unsigned char huge*, and the *farmalloc* system call be used in place of *calloc*.

Another issue to consider is that of pixel depth, or the number of bytes used to represent a pixel, which is dependent on the implementation platform. Typically, digitization produces a byte per pixel quantization, hence pixels have a depth of 1 and are represented as unsigned characters. This has been assumed in this example program and the input and output image structure *Type* variables have been initialized to *BASIC*, or unsigned character. Float or complex pixels may be needed for certain imaging operations, and this demands pixels of larger size. The C macro *sizeof()* can be used to auto-specify the pixel depth independent of platform and we show an example of this at the end of this discussion.

```
/** Basic Image Processing Program **/

#include <stdlib.h>
#include <memory.h>
#include "Image.h"

main()
{
   /* Functions called by main */
   extern void Img_in(struct Image *In),
      Convolve(struct Image *In,
     struct Image *Out, struct Image *Mask),
      Img_out(struct Image *Out);

   /* Declare Input & Output Images */
   struct Image In, Out, Mask;

   /* Indexes for mask generation section */
   signed char *tmp;  int i;

   /* Init image parameters */
   In.Rows = Out.Rows = 256;
   In.Cols = Out.Cols = 256;
   In.Type = Out.Type = BASIC;

   /* Allocate the memory for the images */
   In.Data = (unsigned char *)
             calloc(In.Rows,In.Cols);

   Out.Data = (unsigned char *)
              calloc(Out.Rows,Out.Cols);

   /* Set up a 5x5 Mask */
   Mask.Rows = Mask.Cols = 5;
   Mask.Type = BASIC;
   Mask.Data = (unsigned char *)malloc(25);

   /* Init the 5x5 Mask image as a
      LaPlacian, which looks like:
         -1 -1 -1 -1 -1
         -1 -1 -1 -1 -1
         -1 -1 24 -1 -1
         -1 -1 -1 -1 -1
         -1 -1 -1 -1 -1     */
```

```
    /* set all mask values to -1 */
    tmp = (signed char *)Mask.Data;
    for(i=0; i<25; ++i){
       *tmp = -1;
       ++tmp;
    }

    /* now fix the middle one */
    tmp = (signed char *)Mask.Data + 13;
    *tmp = 24;

    /* Now do the processing */
    Img_in(&In);       /* Input image file */
    Convolve(&In, &Mask, &Out);
    Img_out(&Out);     /* Output the result */
}
```

The input of data involves the opening of files or the accessing of an image hardware device. There are many ways to load image data from a file; the following version of *Img_in* is possibly the simplest. It assumes that the raw image data is stored in a binary file called **MYIMAGE.RAW** in row-column order. The term, *raw*, is typically used to describe digitized data that has had no processing applied. The routine accepts the pointer of an image structure to be read, then loads it from the file a row at a time(*Img->Cols*).

```
/* Input an image file */

void Img_in(struct Image *Img){
    FILE *ifile;
    int i;

    /* open the file for binary reading */
    ifile = fopen("MYIMAGE.RAW", "rb");

    /* read directly into the image array */
    for(i=0; i<Img->Rows; ++i)
       fread(Img->Data + i*Img->Cols,
          Img->Cols,1,ifile);
    fclose(ifile);
}
```

The image processing aspect of the program is performed by the Convolve function as described in the topics section of this book. Here the Input image, *In*, is convolved with an image mask, *Mask*, and the result placed in image *Out*. Note that this is a circular convolution, where the mask is not centered on the image pixel being operated on. The resulting convolution is shifted by N/2 pixels, where N is the size of the mask.

```
/* 2-D Discrete Convolution */

void Convolve(struct Image *In,
      struct Image *Mask, struct Image *Out)
{
   long i,j,m,n,idx,jdx;
   int ms,im,val;
   unsigned char *tmp;

   /* the outer summation loop */
   for(i=0;i<In->Rows;++i)
      for(j=0;j<In->Cols;++j){
         val = 0;
         for(m=0;m<Mask->Rows;++m)
            for(n=0;n<Mask->Cols;++n){
               ms = (signed char)
                     *(Mask->Data +
                       m*Mask->Rows + n);
               idx = i-m;
               jdx = j-n;
               if(idx>=0 && jdx>=0)
                  im = *(In->Data +
                         idx*In->Rows + jdx);
               val += ms*im;
            }
         if(val > 255)val = 255;
         if(val < 0)   val = 0;
         tmp = Out->Data + i*Out->Rows + j;
         *tmp = (unsigned char)val;
      }
}
```

The final step in the program is to output the convolution result. This is accomplished in a similar fashion to the way that the image was input to the program. The file

CONVOUT.RAW is opened for writing with the *fopen* routine and the image data written to it. If the file exists, it is written over, if it does not exist, it is created. In the simplest case the *Img_out* routine is as follows:

```
/* Output an image to a file */

void Img_out(struct Image *Out){

   FILE *ofile;
   int i;

   /* open (or create) a file for writing */
   ofile = fopen("CONVOUT.RAW", "wb");

   /* Output the image by rows */
   for(i=0; i<Out->Rows; ++i)
      fwrite(Out->Data + i*Out->Cols,
         Out->Cols,1,ofile);

   fclose(ofile);
}
```

Notice that the image input and output routines are identical except for the file open call, the filenames, and the use of *fread* or *fwrite* in the loop. It would be a simple matter to combine the read and write into a single call with a flag indicating the direction of data flow. In addition, a character pointer can be passed in the call to provide the file name to open. A more advanced routine incorporating these changes is given in the *Img_IO* code below:

```
/* Input or output image data */

void Img_IO(struct Image *IO,
            char *filename, char dir)
{
   FILE *fp;
   int i;

   /* dir is non-zero for reading */
   if(dir)                        /* READ */
      fp = fopen(filename, "rb");
   else                           /* WRITE */
      fp = fopen(filename, "wb");
```

```
   /* process the image by rows */
   for(i=0; i<IO->Rows; ++i)
      if(dir)
         fread(IO->Data + i*IO->Cols,
                  IO->Cols,1,fp);
      else
         fwrite(IO->Data + i*IO->Cols,
                  IO->Cols,1,fp);

   fclose(fp);
}
```

It is important to understand how to access image files with pixel data of other than type *unsigned char*. For example, complex images consist of float type pixels (generally four bytes per pixel) that are stored sequentially in pairs, the first float pixel being the real component of the complex image and the second pixel being the imaginary component. Below is a routine to read a complex image. Note the casting of the image data pointer to float.

```
/* Input a complex (float) image file */

void Img_in(struct Image *Img,
               char *filename)
{
   FILE *ifile;
   int i,sz;
   unsigned char c;
   float *fptr;

   /* open the file for reading */
   ifile = fopen(filename, "rb");

   fptr = (float *)Img->Data;

   sz = Img->Rows * Img->Cols;

   /* read directly into the image array */
   for(i=0; i<sz; ++i){
      fread(fptr,sizeof(float),2,ifile);
      fptr += 2;
   }

   fclose(ifile);
}
```

From the previous example, the C macro *sizeof()* was used to specify the number of bytes to read of float data. The two pixel components are then read one after the other for a total of *sz* pixels. The image structure data pointer, *fptr*, was incremented by two on each pass through the loop to accommodate the real and imaginary components. It will automatically change by *sizeof(float)* bytes because it was declared as a float . The *Img->Data* pointer is cast from unsigned char to float in the *fptr* assignment statement.

We can now, as a final exercise, rewrite Img_IO so that *any* data type image may be read or written based on the value of the image structure *Type* variable, as shown below:

```
void Img_IO(struct Image *IO,
            char *filename, char dir)
{
   FILE *fp;
   int i,sz_pix;

   if(dir)
      /* open (or create) a file for read */
      fp = fopen(filename, "rb");
   else
      /* open (or create) a file for write */
      fp = fopen(filename, "wb");

   switch(IO->Type){
      case BASIC:
         sz_pix = sizeof(unsigned char);
         break;

      case UINT:
         sz_pix = sizeof(unsigned int);
         break;

      case REAL:
         sz_pix = sizeof(float);
         break;

      case CMPLX:
         sz_pix = 2*sizeof(float);
         break;
   }
```

```
   /* process the image by rows */
   for(i=0; i<IO->Rows; ++i)
      if(dir)
         fread(IO->Data + i*IO->Cols,
                  IO->Cols,sz_pix,fp);
      else
         fwrite(IO->Data + i*IO->Cols,
                  IO->Cols,sz_pix,fp);
   fclose(fp);
}
```

The types and formats of image data structures are highly variable and depend on the hardware that captures them, the computer that processes them, and the terminals that display them. Carefully select the data formats that you use, and if problems or erratic results appear in your programs, look first at your image data representation scheme.

The table below is a partial list of Tags in TIF files. In some cases, for brevity, the Tag names have been abbreviated.

Name	Tag	†	Count
NewSubfileType	254	L	1
SubfileType	255	S	1
ImageWidth	256	S	1
ImageLength	257	S	1
BitsPerSample	258	S	SamplesPerPixel
Compression	259	S	1
PhotometricInterpretatn	262	S	1
Thresholding	263	S	1
CellWidth	264	S	1
CellLength	265	S	1
FillOrder	266	S	1
DocumentName	269	A	n/a
ImageDescription	270	A	n/a
Make	271	A	n/a
Model	272	A	n/a
StripOffsets	273	S	StripsPerImage
Orientation	274	S	1
SamplesPerPixel	277	S	1
RowsPerStrip	278	S	1
StripByteCounts	279	L	StripsPerImage
MinSampleValue	280	S	1
MaxSampleVal	281	S	1
XResolution	282	R	1
YResolution	283	R	1
PlanarConfiguration	284	S	1
PageName	285	A	n/a
XPosition	286	R	n/a
YPosition	287	R	n/a
FreeOffsets	288	L	
FreeByteCounts	289	L	
GrayResponseUnit	290	S	1
GrayResponseCurve	291	S	2**BitsPerSample
Group3Options	292	L	1
Group4Options	293	L	1
ResolutionUnit	296	S	1
PageNumber	297	S	2

ColorResponseCurve	301	S	3*(2**BitsPerSample)
Software	305	A	n/a
DateTime	306	A	20
Artist	315	A	n/a
HostComputer	316	A	n/a
Predictor	317	S	1
WhitePoint	318	R	2
PrimaryChromaticities	319	R	6
ColorMap	320	S	3*(2**BitsPerSample)

† -- Tag Type:
A=ASCII, B=BYTE, S=SHORT,
L=LONG, R=RATIONAL.

Although the Aldus and Microsoft Corporations do not have formal commitments to the maintenance of a TIF standard, they are receptive to questions and issues regarding TIF, or to requests for Tag assignments. They may be contacted at:

Developers Desk
Aldus Corporation
411 First Ave. South
Suite 200
Seattle, WA 98104
(206) 622-5500

Windows Marketing Group
Microsoft Corporation
16011 NE 36th Way
Box 97017
Redmond, WA 98073-9717
(206) 882-8080

Algorithms in C, Sedgewick, R., Addison-Wesley, 1990.

Bit-Mapped Graphics, Rimmer, S., Windcrest Books, 1990.

Computer and Robot Vision, Volumes I & II, Haralick, R. M. and Shapiro, L. G., Addison-Wesley, 1992.

Computer Imaging Recipes in C, Myler, H. R. and Weeks, A. R., Prentice Hall, 1993.

Computer Image Processing and Recognition, Hall, E. L., Academic Press, 1979.

Computer Vision, Ballard, D. H. and Brown, C. M., Prentice Hall, 1982.

Digital Image Processing, 2nd Ed., Gonzalez, R. C. and Wintz, P., Addison-Wesley, 1987.

Digital Image Processing, Pratt, W. K., Wiley, 1978.

Digital Image Processing and Computer Vision: An Introduction to Theory and Implementation, Schalkoff, R. J., Wiley, 1989.

Digital Picture Processing, Volumes I & II, Rosenfeld, A. and Kak, A. C., Academic Press, 1982.

The Image Processing Handbook, Russ, J. C., CRC Press, 1992.

Machine Perception, Nevatia, R., Prentice Hall, 1982.

Machine Vision and Digital Image Processing Fundamentals, Galbiati, L. J., Prentice Hall, 1990.

Nonlinear Digital Filters, Pitas, I. and Venetsanopoulous, A. N., Kluwer Academic, 1990.

Numerical Recipes in C, Press, W.H., Flannery, B.P., Teukolsky, S.A. and Vettering, W.T., Cambridge University Press, 1988.

adaptive filters -- filters that change their characteristics as they are applied to an image.

adaptive window edge detected filter -- this filter adaptively changes its window size when an edge is detected to preserve the details of the edge.

aperture -- size of a lens opening, often controlled with a metal iris.

array processor -- specialized computer designed to perform calculations on arrays (or images) rapidly.

aspect ratio -- ratio of height to width of an object either captured by a camera or displayed by a monitor.

autoscaling -- alogrithm that scales an image between a minimum and a maximum gray level value following the application of an algorithm that modifies graylevel.

binary image -- image where pixels have only two values, generally 0 and 1.

brightness -- the gray level value of a pixel within an image that corresponds to energy intensity. The larger the gray level value the greater the brightness.

Cathode Ray Tube (CRT) -- electronic tube that allows display of images and graphics through the electronic positioning of an electron beam; the glass screen of a computer display or monitor.

CCD camera -- solid-state camera using a charge-coupled device sensor.

C-mount -- common lens mounting system used on electronic cameras, 1" in diameter with 32 threads/inch.

closing -- a morphological operation that smooths the geometrical contour of objects within an image. This operation is composed of a morphological dilation operation followed by a morphological erosion operation.

complex numbers -- number system represented by the sum of a pair of real values **a** and **b**, written **a** +**b***j*, where **a** is called the real part(sometimes Re) and **b** the imaginary part (sometimes Im). The term imaginary is used because the second value, **b**, is multiplied by the imaginary operator $j=\sqrt{-1}$.This is simply a convention that allows easy representation of frequency dependent functions. The usefulness of complex numbers is revealed when *phase* and *magnitude spectrums* are derived from the results of Fourier transformations.

composite video -- RS-170 video that includes synchronization signals.

control point -- corresponding points selected between two images so that they can be aligned to eah other using a warping process.

contrast -- the amount of gray level variation within an image

convolution -- See *discrete convolution.*

convolution mask -- small subimage, typically 3x3 to 7x7 in size, used as a filter in a discrete convoultion operation. Examples of convolution masks are:

$$\text{3x3 Low Pass (smoothing) Filter}\begin{bmatrix} 1 & 2 & 1 \\ 2 & 4 & 2 \\ 1 & 2 & 1 \end{bmatrix}$$

$$\text{5x5 High Pass (sharpening) Filter}\begin{bmatrix} 0 & -1 & 1 & -1 & 0 \\ -1 & 2 & -4 & 2 & -1 \\ -1 & -4 & 13 & -4 & -1 \\ -1 & 2 & -4 & 2 & -1 \\ 0 & -1 & 1 & -1 & 0 \end{bmatrix}$$

See *discrete convolution.*

cornea -- transparent outer surface of the eye that performs the initial focussing process.

cursor -- graphic object used in display systems to identify the location of a pointing device, such as a mouse, joystick or digitizing pad.

cutting & pasting -- process of outlining an area in an image, removing it (cutting) or adding it (pasting) to either the same image or a different one.

depth of field -- twice the distance an object in focus may move from the object plane and still remain in focus.

digitizer -- electronic circuit that converts analog, or continuous signals into discrete or digital data.

dilation (⊕) -- a morphological operation that *enlarges* the geometrical size of objects within an image.

dither -- term used to describe computer algorithms that simulate grayscale output on binary devices; see *half-toning*.

discrete -- refers to signals or data that is divided into samples, or fixed quantities.

Discrete Cosine Transform -- mathematical transformation performed on discrete data that resolves additive real sinusoidal components of the data that correspond to the spatial frequency content of the data.

discrete convolution -- process where two images are combined using a shift, multiply and add operation. Typically, one image is substantially smaller than the other and is called the *mask* or window. Masks can be designed to perform a wide range of filtering functions. See *mask*, *spatial filter*.

Discrete Fourier Transform -- mathematical transformation performed on discrete data that resolves additive complex sinusoidal components of the data that correspond to the spatial frequency content of the data.

double window modified trimmed mean adaptive filter -- filter that uses two windows that adaptively switch from a 3 by 3 *median* to a 5 by 5 *mean* filter.

electromagnetic spectrum -- range of known energy wavelengths (or frequencies) and their corresponding labels.

enhancement -- algorithms and processes that improve an image based on subjective measures.

erosion (Θ) -- morphological operation that *reduces* the geometrical size of objects within an image.

f-number -- aperture setting of a lens; ratio of the diameter of the aperture to the focal length of the lens; the f is an abbreviation for *field*, not focal length, as the f-number determines the Depth of Field.

f-stop -- see *f-number*.

Fast Fourier Transform (FFT) -- a special formulation of the Fourier Transform (see *Discrete Fourier Transform*) that takes advantage of repetitive forms to increase the speed of computer calculations.

focal length -- point at which the rays converged by a lens meet; may be changed by the aperture setting, see *f-number*.

frame -- term used to describe an image, typically in context with a series of images, such as a single frame in an image sequence.

frame-buffer -- computer memory designed to store an image or set of images that have been captured and digitized by a *frame-grabber*, or *digitizer*.

frame-grabber -- electronic circuit that converts (using digitization) an analog video signal into a digital image, see *digitizer*.

frequency -- measure of periodicity of a data set, or how often the data repeats a pattern in a given measure, such as time or distance. See *periodic, spatial frequency*.

gamma -- basic measure of contrast; in film terminology, gamma is the slope of the density vs exposure curve; in electronic display terminology, gamma is the slope of the brightness distribution curve; large gamma indicates a steep slope and high contrast.

ganglia -- grouping of nerves in the retina that combine the signals from the light sensory nerves giving rise to a low-pass filtering effect.

gas-plasma display -- display that uses the electroluminescence of rare gases to create visible output for a computer monitor.

gaussian noise -- a type of noise whose histogram is Gaussian (bell) shaped.

graphic tablet -- computer input device that transforms pen position on a surface into position coordinates.

Graphic Interchange Format©(GIF) -- file storage format for images developed by Compuserve Information Service, Inc.; uses LZW compression.

graylevel -- value of gray from a black and white (monochrome) image.

grayscale -- range of gray shades, or graylevels, corresponding to pixel values that a monochrome image incorporates.

Hadamard Transform -- transform that resolves a data set into sets of square waves, where the maximum value is 1 and the minimum value is -1. Sometimes called the Walsh-Hadamard transform (see *Walsh Transform*), the Hadamard Transform is distinguished from the Walsh in that the transform matrices may be generated recursively using the lowest order Hadamard matrix,

$$H_2 = \begin{bmatrix} 1 & 1 \\ 1 & -1 \end{bmatrix}$$

and the recursive matrix,

$$H_{2N} = \begin{bmatrix} H_N & H_N \\ H_N & -H_N \end{bmatrix}.$$

The H_{2N} matrix is the Hadamard matrix of order 2N and higher orders are easily generated by applying the recursive relation shown above.

half-toning -- technique for rendering a grayscale effect on a binary (two-tone) output device.

histogram -- distribution of pixel graylevel values. A graph of number of pixels at each graylevel possible in an image. A histogram is a probability distribution of pixels values and may be processed using statistical techniques. These processes result in changes to the brightness and contrast in an image, but are independent of the spatial distribution of the pixels. See *uniform histogram, histogram stretching, histogram equalization, histogram specification.*

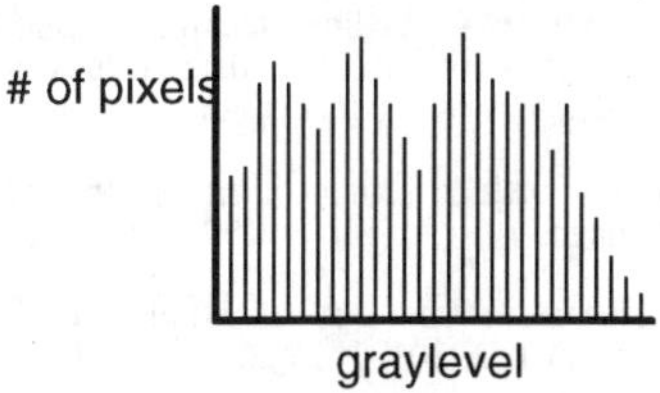

histogram equalization -- process that converts an images histogram to a uniform distribution. This is accomplished by integrating (summing) the histogram over all graylevel values. The effect of equalization is improved contrast in the image.

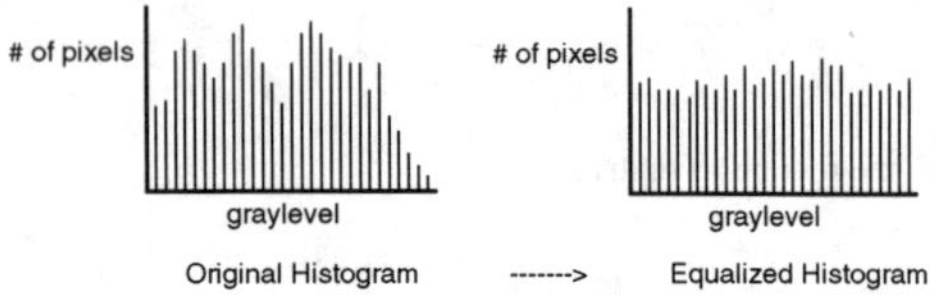

histogram specification -- process that changes the shape of a given image histogram to that of another, specified by the user. The process is used when the histogram of one image is desired in another, or during an interactive histogram modification scheme where the user is allowed to change the histogram dynamically to acheive a desired contrast result.

histogram stretching -- process that scales a histogram to the fullest possible range. This is distinguished from histogram equalization, which is the conversion of a histogram to a uniform distribution.

homomorphic filter -- filter that uses logarithm to seperate intensity and reflection components of an image so that each can be modified independently.

Huffman Coding -- coding technique that calculates probability of occurrance for data values and assigns smallest codes to most frequent data.

illumination -- outside source of energy that illuminates a scene or image.

integer numbers -- the set of whole numbers of the following form 1, 2, 301, 1024 etc.

interlacing -- process of skipping every other line in a output or input scheme.

intersection -- the overlapping region of two objects or sets.

Lempel-Ziv-Welch (LZW) Coding -- coding scheme similar to Huffman (see above) where probabilities are recalculated when performance changes.

lens -- transparent device used to bend and focus light rays.

liquid-crystal display -- display that uses the light attentuating effect of amorphous crystals to create visible output for a computer monitor.

magnification -- ratio of the size of an objects image to the actual size of the object.

magnitude spectrum -- spectrum of spatial frequency magnitudes (or strengths) in an image. The *magnitude spectrum* is an image where each pixel represents the magnitude of the spatial frequency at that location from the original image. The spectrum is derived from a complex spatial frequency generating transform, such as the Discrete Fourier. The magnitude(A) at a given pixel location is given by the equation, $A = \sqrt{Re^2+Im^2}$, where Im is the imaginary portion of the complex transform and Re is the real portion. See *complex numbers*, *frequency*, *spectrum.*

mask -- generally refers to a small image used to specify the area of operation to take place on a larger image in an algorithm. Mask also refers to a discrete convolution filter. See *convoultion mask.*

maximum filter -- this filter replaces the pixel being operated on with the maximum graylevel of a set pixels located under a spatial mask.

mean -- the average of a set of data values; e.g., for the set of values:

$$\{ 3, 5, 6, 7, 4, 3, 2, 2 \}$$

the mean is $(3+5+6+7+4+3+2+2)/8 = 4$.

medial axis transform -- the *skeleton* of an object.

median -- the middle value of a set of *ordered* data values; e.g., for the set of values:

$$\{ 3, 5, 6, 7, 4, 3, 2, 2 \}$$

the median is {2 2 3 3 | 4 5 6 7} = 3.5

mensuration -- measurement algorithms.

minimum filter -- this filter replaces the pixel being operated on with the minimum gray level of a set pixels located under a spatial mask. See *maximum filter*.

monochrome -- literally *one color*, also used to describe black and white grayscale images.

monitor -- display used to output images or computer data.

mouse -- computer input device that is moved on a surface and translates physical movement into position data.

multi-level threshold -- the process of thresholding the graylevel values of an image into multiple levels.

negative exponential noise -- a type of noise that is described by a negative exponential histogram.

NTSC -- acronym for National Television Standards Committee; term used to describe RS-170 compatible color video.

nyquist theorem -- sampling theorem that requires that a signal be sampled, or *digitized*, at a rate (the nyquist rate) that is twice the highest frequency present in the sampled signal. When the nyquist rate is used, all components of the sampled signal will be adequately represented.

null set -- a set of objects containing no members.

opening -- morpholgical operation that is used to smooth the geometrical shape of objects within an image. Opening is a morphological erosion followed by a morpholgical dilation operation.

optic nerve -- nerve that carries image data from the eye to the brain.

optimum threshold -- this is the best threshold value for a particular image to reveal the most possible objects.

order statistics -- the process of ordering a set of data from minimum to maximum to obtain a set of statistics.

outliers -- pixels that contain gray level values that do not represent the normal gray level value within a region of an image.

outline -- the contour of objects within an image.

pel -- European term for pixel.

periodic -- when a data set, or signal, repeats itself, the data is said to be *periodic*. The size of the data subset (typically measured in time or distance) that repeats itself is called the *period* of the data set. *Frequency* is the measure of periodicity and is given as periods/time or periods/distance.

phase spectrum -- spectrum of spatial frequency phase (or directions) in an image. The *phase spectrum* is an image where each pixel represents the phase of the spatial frequency at that location from the original image. The spectrum is derived from a complex spatial frequency generating transform, such as the Discrete Fourier. The phase(Q) angle at a given pixel location is given by the equation, $Q = \tan^{-1}(Im/Re)$, where Im is the imaginary portion of the complex transform and Re is the real portion. See *complex numbers, frequency, spectrum.*

pixel -- slang for picture element, the smallest element of an image; pixels are arranged in row and columns to create an image, frame or picture.

pupil -- aperture of the eye; term used for variable aperture.

profile -- imaging function that plots or displays pixel data along a line within an image to yield a cross section of values.

quantization -- range of values that a pixel can represent.

real numbers -- a set number of the form 3.12, 4.518, 6.323 etc.

real-time -- 30 frames per second, or the number of frames required so that normal motion is not blurred to a human observer.

reconstruction -- algorithms and processes that attempt to construct a two-dimensional image from one-dimensional data functions (CAT scans, Synthetic Aperature Radar, etc.).

reflectance -- portion of incident light that is relected from objects or background in an image.

resolution -- smallest feature (spatial) or graylevel value (quantization) that an image system can resolve.

restoration -- algorithms and processes that attempt to remove a degradation (noise, blurring and defocussing effects) based on an objective criterion.

retina -- spatial light sensor array of the eye.

rubber-band -- when a graphic selection contour or object stretches to follow the input cursor.

RF modulator -- device that converts an RS-170 video signal into a radio frequency signal so that a television can receive it through its tuner.

RS-170 -- video standard of 525 lines, interlaced at 1/30 second.

Run-Length Encoding (RLE) -- simple coding scheme consisting of number pairs where one number represents a pixel value and the other the number of times the value is repeated.

salt and pepper noise -- noise that contains both minimum and maximum outlier pixels. In a 256 gray level image, the pepper noise has gray level value of 0, while the salt noise has a gray level value 255.

sampling -- used to describe spatial resolution of an image.

sawtooth wave -- used within the slant transform

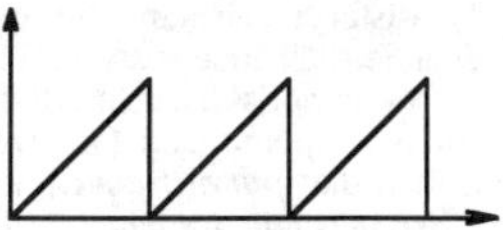

separability -- two-dimensional (image) transform property where the mathematical operations defining the transform can be divided into two or more parts. This property is advantageous from a computations standpoint. The *Fourier*, *Cosine*, *Walsh* and *Hadamard Transforms* are seperable.

set -- a collection of objects combined together that all contain something in common.

signal adaptive median filter -- An adaptive median filter that changes its window size and characteristics depending on the input signal/image.

skeletonization -- algorithm used to find the central axis (skeleton) of an image object.

Sobel -- directional edge detection discrete convolution masks of the following form:

$$\text{vertical}\begin{bmatrix} -1 & 0 & 1 \\ -2 & 0 & -2 \\ -1 & 0 & 1 \end{bmatrix} \quad \text{horizontal}\begin{bmatrix} -1 & -2 & -1 \\ 0 & 0 & 0 \\ 1 & 2 & 1 \end{bmatrix}$$

solid-state cameras -- electronic cameras that use solid-state arrays as their sensing element, see *CCD camera*.

spatial filter -- image filter that operates on the spatial distribution of pixel values in a small neighborhood. Although a *spatial frequency* filter operates on spatial distributions of pixels, this term is generally reserved for *discrete convolutions* while the latter term is used for filters derived from image transforms, See *discrete convolution.*

spatial frequency -- measure of the periodicity of a data set with respect to a distance measure. Periodic changes in brightness values across an image are defined in terms of spatial frequency, or periods/distance. If the period of a brightness pattern is 20 pixels and the size of a pixel is 1/20th of a mm, then the *spatial frequency* of the data set is 20 pixels/mm. See *periodic, freqency.*

spectrum -- a collection of ordered frequencies describing the frequency content of a data set, or signal. For example, the *electromagnetc spectrum* is the collection of light frequencies that are physically known to science. See *electromagnetc spectrum, magnitude spectrum, phase spectrum.*

square wave -- a rectangularly shaped signal ().

standard image -- 512 by 512 pixels, 8-bits (1 byte) quantization per pixel.

structuring set -- the set of pixels used to describe the structuring function used in the morphological erosion and dilation operations.

Tagged Interchange File Format (TIFF) -- image storage file format developed by Microsoft and Aldus corporations; most commonly used in desktop publishing applications and with image scanner hardware.

template -- subimage used in image correlation or matching function. Sometimes used to describe discrete convolution mask. See *discrete convolution.*

threshold -- a value used to segment the graylevel values of an image into two different regions. Also called the binarization of an image. For example, if a threshold value of 128 is chosen, then any pixel below this value would be set to 0 and all pixels greater than and equal to this value would be set to 255.

touch-screen -- computer input device that reports the coordinates of where the monitor screen was touched either by a finger or wand.

trackball -- computer input device that moves the cursor based on position of a ball mounted to free rotate in a fixture. As the ball is rotated, the cursor tracks its motion in the same direction.

uniform noise -- a type of noise described by a uniform histogram.

union -- the process of combining two different sets into one set.

variance -- the average value that a set of data values differ from the *mean* of the set. Formally, it is the average value of the *squares* of the deviations from the mean; e.g., for the set of values:

{ 3, 5, 6, 7, 4, 3, 2, 2 }

the mean is 4 , the data set of *differences* from the mean is:

{ -1, -1, 2, 3, 0, -1, -2, -2},

the data set of *squares* of the *differences* is:

{ 1, 1, 4, 9, 0, 1, 4, 4},

and the *mean* of this set, the *variance*, is 3.

See *mean*..

Venn diagrams -- a graphical method of performing various set operations such as union (or) and intersection (and).

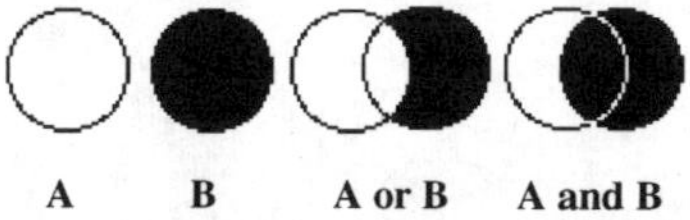

video -- signal that carries analog image information.

video output controller -- subunit of an image processing system that controls and routes video signals in the system.

vidicon -- electronic tube that images using a scanned electron beam.

visible spectrum -- portion of the electromagnetic spectrum that is visible to the human eye; color spectrum; see *electromagnetic spectrum.*

WYSIWYG -- What You See Is What You Get; term used to describe computer displays that show what text and graphics will appear when a document is printed.

Walsh Transform -- transform that resolves a data set into sets of square waves, where the maximum value is 1 and the minimum value is -1. Sometimes called the Walsh-Hadamard transform (see *Hadamard Transform*), the Walsh Transform is distinguished from the Hadamard in that the transform has a fast form implementation similar to that used by the *Fast Foruier Transform.* The Walsh-Hadamard transforms have substantial advantages in computation over other transforms in that no complex or floating-point numbers are required.

warp -- to perform a geometric distortion operation on an image using a computer algorithm.

weber ratio -- ratio of background intensity to foreground intensity in human visual perception; this ratio remains close to 20% over a large range of brightness values.

zoom -- process by which an image is magnified by a computer algorithm.

A

B

C

G

H

I

N

O

P

Notes

Notes

Notes

Notes

Notes